# THE EVERYTHING® GUIDE TO ANATOMY AND PHYSIOLOGY

Dear Reader,

Thank you for reading *The Everything® Guide to Anatomy and Physiology*. This was always one of my favorite subjects when I was getting my biology degree because it was the first subject that really applied to something I related to every day: my own body. Since the human body is such a complex and intricate machine, I hope you find this guide a simplified view of the more challenging designs and mechanics that keep our bodies running smoothly. Too often textbooks, which are written by professors who deal with research on explicit detail, are written in technical language and assume a certain level of understanding of the biological processes. As a student I found this frustrating and confusing. As you learn about the body and begin to understand its beautiful pattern, design, and function, use this guide as a supplement to assist you through those most challenging portions of human anatomy and physiology. This information will be essential to those in the biological sciences as well as those wishing to pursue a career in the health professions. I wish you the best of luck in all your studies of the amazing human body.

Kevin Langford, PhD

## Welcome to the EVERYTHING® Series!

These handy, accessible books give you all you need to tackle a difficult project, gain a new hobby, comprehend a fascinating topic, prepare for an exam, or even brush up on something you learned back in school but have since forgotten.

You can choose to read an Everything® book from cover to cover or just pick out the information you want from our four useful boxes: e-questions, e-facts, e-alerts, and e-ssentials.

We give you everything you need to know on the subject, but throw in a lot of fun stuff along the way, too.

We now have more than 400 Everything® books in print, spanning such wide-ranging categories as weddings, pregnancy, cooking, music instruction, foreign language, crafts, pets, New Age, and so much more. When you're done reading them all, you can finally say you know Everything®!

**QUESTION**

Answers to common questions

**FACT**

Important snippets of information

**ALERT**

Urgent warnings

**ESSENTIAL**

Quick handy tips

**PUBLISHER** Karen Cooper

**MANAGING EDITOR, EVERYTHING® SERIES** Lisa Laing

**COPY CHIEF** Casey Ebert

**ASSISTANT PRODUCTION EDITOR** Alex Guarco

**ACQUISITIONS EDITOR** Lisa Laing

**ASSOCIATE DEVELOPMENT EDITOR** Eileen Mullan

**EVERYTHING® SERIES COVER DESIGNER** Erin Alexander

*I dedicate this guide to all students of anatomy and physiology past, present, and future, and to those amazing scientists who continue to bring to light the intricacies of the human body.*

Adams Media
An Imprint of Simon & Schuster, Inc.
57 Littlefield Street
Avon, Massachusetts 02322


Manufactured in the United States of America

10 9 8 7 6 5 4 3 2

Library of Congress Cataloging-in-Publication Data has been applied for.

ISBN 978-1-4405-8182-3
ISBN 978-1-4405-8183-0 (ebook)

# THE EVERYTHING® GUIDE TO ANATOMY AND PHYSIOLOGY

All you need to know about how the human body works

Kevin Langford, PhD

Adams Media
New York London Toronto Sydney New Delhi

# Contents

# Acknowledgments

I would like to acknowledge the most influential people from my education and training who have enabled me to be a part of this great study of the human body. Dr. Don Hay taught me during many of my courses in human biology, as well as initially trained me in research. Dr. Ralph Sanderson showed me by example how to study a problem and how to communicate effectively. Dr. John Moore introduced me to the joy of academic writing. He is a great mentor and has become a valued friend.

Lastly, James Langford exemplified all those things that make a person great. He was my first "favorite" and best teacher. He still demonstrates an amazing work ethic for those things that are important and deserve your best efforts. His caring nature and servant's heart help me be a better professor and advisor to my students. He is my dad, my best friend, and my example. To him I owe eternal thanks and gratitude.

## Top Ten Ways You'll Use Your Knowledge of Anatomy and Physiology in Everyday Life

1. In the gym: You'll know which exercises work each muscle (and later, which ones are hurting the most).
2. In the doctor's office: You'll know the functions of the organs your doctor is talking about.
3. At your local trivia night: You'll be the champion of science trivia.
4. To help friends and family members: You can be the local anatomy expert.
5. To stay healthy: Not only will you be able to eat better and take better care of your body, but you will know when and how to avoid sick people.
6. If you're preparing for a trip to Mars: Counteracting the changes to the human anatomy and physiology during long-term space travel will be essential for a trip to Mars (will you be the first?).
7. To make some extra cash: Become a teaching assistant or AP tutor.
8. To create a health and fitness app: Health and fitness are increasingly important to everyone. Will you create the next Fitbit?
9. On a diving vacation: Understanding the physiology of pressure and blood gases is critical to your survival when scuba diving.
10. For your career: If you want to have any type of medical career, you'll need a good understanding of human anatomy and physiology.

# Introduction

HUMAN ANATOMY AND PHYSIOLOGY are the fascinating studies of human form and function. In many ways, the form is the function. However, cells, tissues, and organs are often intricately organized to facilitate many functions simultaneously. In this book, all of the processes and structures of the human body will be explained. (Many of the complex biochemical processes will be simplified for easier understanding.)

The human body has always amazed mankind. Early scientific drawings and diagrams demonstrate the long-standing fascination with the body. Even cave drawings and later hieroglyphics illustrate that man was aware of the complex machine that was his body. In many ways, you may very well view the body as an overly complex device, and feel overwhelmed at the amount of detail that must be absorbed when studying it. However, by starting with a solid foundation on which to build a more detailed understanding, anyone can master the human body.

Understanding the body one system at a time is an important part of the organization of this book. This organization will help you gain knowledge slowly and effectively. At the same time, it is also important not to lose sight of how each system works relative to the overall view of the body: the big picture. This is not unlike putting together a puzzle. The picture on the front of the box is the guide for completion; however, individual pieces are placed together based on similarities in color, design, and shape into a small portion of the puzzle that you will build upon.

As with most of the biological sciences, the deductive approach to understanding anatomy and physiology is one of the most exciting aspects of learning about this topic. You are already intimately familiar with the sensations and many of the workings of your body; this book will explain their causes and purposes to you.

In *The Everything® Guide to Anatomy and Physiology*, you will find examples that will bring these points into focus and will build upon your understanding and experience with the human body. You will also find many real-life examples in this book that will simplify your understanding of a myriad of intricate biochemical processes or complex biological designs. You will also find many questions that will cause you to think about this topic from a simple, but important, perspective.

This book is *not* meant to be a replacement for a textbook. Rather, it is another way you can master and attain the complex information presented in an anatomy and physiology course. You can also use this resource to get a fresh look at the subject if you are struggling to fully understand any information presented in a course or other book.

CHAPTER 1

# The Chemistry of Cells

Everything in the universe, from the largest of stars to the smallest atom in the human body, owes its very existence to chemistry. This isn't a shameless plug to recruit chemistry majors; however, it does give credit where credit is due. The interaction of atoms has created the human body and the world it inhabits. Elements are the foundations of cell formation and a cell's functional subunits (called organelles). The variety of elements found in cells allows for greater diversity in biochemical processes, cell types, and cell activities that are essential for the health and survival of an organism.

# The Most Important Elements

Just as the human body doesn't have a single "most important" organ, several elements are essential for the creation of life. In the following section, you'll find information on those elements and their chemical significance (as well as physiological important).

An element is the smallest foundational part of matter, which can't be broken down further without changing it into something different. (i.e., lead can't be changed into gold).

## Elements

Hydrogen (H), carbon (C), nitrogen (N), and oxygen (O) are among the most important elements of all living things on earth. Whether in the air humans breathe, the food humans eat, or the materials that make up the physical structures of the human body, without these elements humanity would not exist. What makes these elements so essential to the formation of life is their ability to interact with other elements and then organize them into important molecules (composed of more than one atom) or compounds (molecules comprising 2 or more elements). They can do this because of their subatomic structure and particles.

## Subatomic Particles

All atoms are made up of 3 basic **subatomic** (i.e., anything smaller than an atom) **particles**: protons, neutrons, and electrons. The number and organization of these particles dictate whether an atom will readily interact with any other atom.

Positively charged protons are found in the nucleus of an atom. The number of these particles present will be reported as the **atomic number** for that element. For example, carbon has an atomic number of 6 and oxygen has an atomic number of 8, which means carbon has 6 protons and oxygen has 8 protons in the nucleus.

Another particle found in the nucleus is the neutron. While neutrons don't contribute any charge to the atom, they do contribute to the mass of the atom. Therefore, the **atomic mass** of an atom is the number of protons *and* neutrons present in the atom. So while carbon has an atomic number of 6 (6 protons), it has an atomic mass of 12 (which means there are also 6 neutrons in the nucleus).

However, while the nucleus is populated, there is an unequal charge for the atom. Like most naturally occurring things in the universe, atoms need balance. Negatively charged particles that orbit around the nucleus balance out the charge of the atom. These are called electrons. It is the electrostatic attraction between the electrons and the protons that keep the electrons spinning in orbit around the nucleus, much like the moon is held close to the earth by gravity. In fact, to obtain a natural balance, atoms need to have the same number of protons as electrons, leaving the atom with an overall net neutral charge.

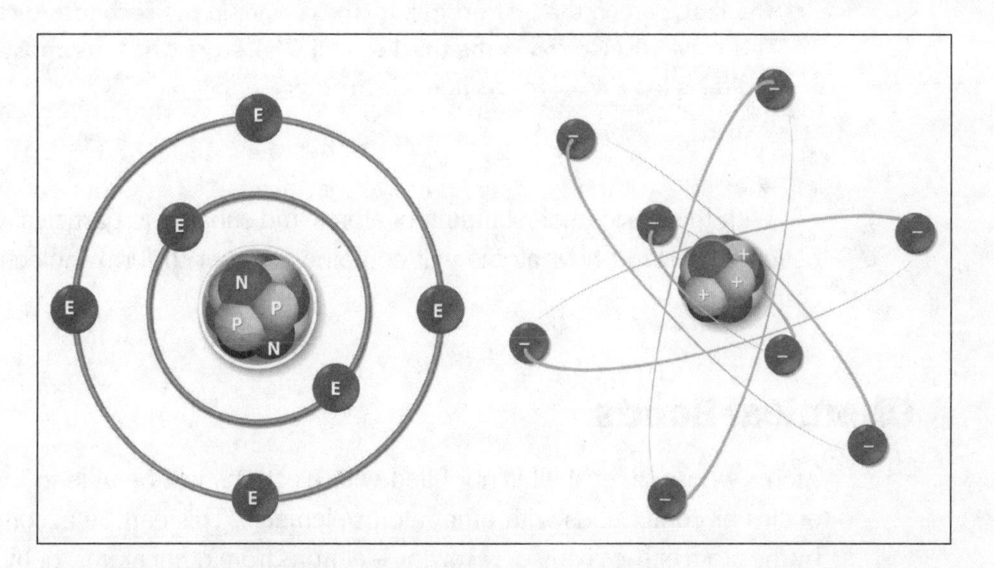


Left: A carbon atom with 2 electrons filling the inner orbital nearest the atomic nucleus and the 4 remaining electrons in the second orbital. Right: An oxygen atom with 8 electrons.

Electrons, however, will not be restricted to a single location. In fact, they are segregated into orbitals or shells around the nucleus. In illustrations, these will often be drawn as concentric circles, with the first being closest to the nucleus and the number of the shells increasing for the ones farther away. For example, up to 2 electrons (the first one or two) will orbit the nucleus in the first orbital closest to the nucleus. After this orbital is filled, if the atom has more electrons, they will be packed into the next orbital. From the second orbital on, the electron capacity is 8 electrons per orbital. Therefore, for carbon, with an atomic number of 6 (meaning 6 protons and thus 6 electrons), 2 of the electrons will be in the first orbital, closest to the nucleus, and the remaining 4 will be in the second orbital. Since carbon has no more than 6 electrons, the second orbital is the outermost orbital for the carbon atom. However, for elements that have more electrons, the packaging of electrons will continue until the final electron has been placed into an unfilled orbital.

**How many electrons are in the outermost orbital of sodium (Na), which has an atomic number of 11?**
The atomic number means there are 11 protons (and 11 electrons). The first 2 are in the first orbital, the next 8 are in the second orbital. This leaves 1 electron in the third orbital. Therefore, the outermost orbital is the third with a single electron present.

With this basic understanding of atoms and subatomic particles, you can better understand how atoms will combine together to form molecules and compounds.

## Chemical Bonds

Atoms whose outer shell is not filled with electrons will be able to form molecules or compounds with other atoms/elements. This can be accomplished by the atom either giving or receiving electrons from other atoms, or by sharing electrons with other atoms.

### Ionic Bond

An **ionic bond** is where 2 atoms form molecules by giving up electrons to others or taking electrons from others to complete their outermost orbitals. The classic example is the compound salt (sodium chloride, NaCl). As mentioned in the sidebar, Na has a single electron in the outermost (third) orbital. To fill the outermost orbital, sodium could recruit 7 more electrons from other atoms, but that would be a lot of work and impractical. Therefore, Na gives up the single electron and leaves the complete second orbital filled with 8 electrons, a very stable arrangement. However, now this atom has 10 electrons and 11 protons. This imbalance between protons and electrons yields an **ion**. An ion is a charged atom that has an unequal number of electrons and protons. In this case, the sodium ion with 10 electrons has an overall positive charge. Chlorine (Cl) has the dilemma of needing a single electron to complete its outermost shell. With an atomic number of 17, there are 7 electrons in the third orbital of chlorine. This is a natural partner for sodium. Sodium gives up its electron to chlorine, which then uses the electron to complete its shell. Since it now has 1 more electron than proton, it has become a chloride ion with an overall negative charge.

This is where the bond happens. The positive charge of the $Na^+$ ion is attracted to the negative charge of the $Cl^-$ ion, and the two will form a moderately strong chemical bond to create NaCl, or salt.

## Hydrogen Bond

Hydrogen bonds are formed when atoms share electrons unequally in compounds. Water is the classical example of this type of bonding. Hydrogen has an atomic number of 1, containing a single electron, so its shell is half full. Oxygen, with an atomic number of 8, is 2 electrons short of filling its outermost shell. Therefore, oxygen will share an electron with 2 hydrogen atoms, which will complete the outer shells of all 3 members of this compound. However, with more protons in the nucleus of oxygen, the shared electrons will be drawn to spend more time around that nucleus than around either hydrogen nucleus. This imbalance will create a slight negative charge on the oxygen side and slight positive charge on the hydrogen arms. This polarization of charge will cause water molecules to be attracted to each other via the opposite charges (e.g., the oxygen will be attached to the hydrogen of another water molecule). In this way, water will adhere to itself and help to create surface tension in collections of water. This type of bond is the weakest of the 3 chemical bonds. It is also the type of bond that holds 2 strands of DNA together in chromosomes (see Chapter 2 for more on chromosomes).

## Covalent Bond

The strongest of the chemical bonds, the **covalent bond**, is formed when a molecule or compound shares electrons equally. Carbon, the foundation atom of organic molecules, is adept at this type of bond formation since its atomic number of 6 means it needs 4 electrons to fill its outermost shell. Because of this, carbon can form 4 single covalent bonds with other atoms. This is clearly illustrated in the basic structure of an **amino acid**, the building block of protein. The carbon is the central atom onto which 4 groups can attach: a carbon (of a carboxyl group, COOH), a nitrogen (of an amine group, $NH_2$), a single hydrogen, and finally a group of variable atoms are arranged into what is termed the R group. The R group is the only portion of the amino acid that will differ among amino acids. (Amino acids will be covered in more detail later in this chapter.)

# pH: Measuring Acids and Bases

The quantity of hydrogen ions ($H^+$) in a solution is determined by the measure of pH of the mixture. If a solution has molecules or compounds that will yield a large number of $H^+$ ions, the solution, based on the mathematical algorithm, will result in a lower pH number and be considered an acid or acidic solution (pH < 7.0). Conversely, with lower $H^+$ concentrations (or higher hydroxyl ions, OH), the pH will be above 7 and considered a base, a basic solution, or alkaline. This standard is set internationally using known materials, such as pure water (pH of 7.0). Biochemical processes are optimized to function in a narrow pH range. Thus, the survival of cells and of organisms depends on the correct pH.

The pH of plasma and body fluids is approximately 7.3–7.4 and is called the physiological neutral point. An acid has a pH of less than 7.0 and a basic solution is higher than 7.0.

# Important Inorganic Compounds

While most biology and biochemical texts will focus on the structure, function, and metabolism (the processing of organic molecules to store or release energy) of **organic molecules** (i.e., molecules containing carbon), some inorganic compounds are essential for human existence and for life in general.

## Water

While most life on the planet is carbon-based, water (which is not carbon-based) is a compound without which life would not be possible. The human body, for instance, is considered to be composed of between 50 and 65 percent water. This water exists within your cells (about ⅔ of your water content), with the remainder outside of the cells in your tissues or blood stream.

Water is the universal solvent because it is composed of **polar molecules**, or molecules that have a positive charge at one end and a negative charge at the opposite end, much like the opposite poles of a battery. Polar molecules are capable of ionizing many other molecules (e.g., NaCl). It is also an essential **substrate** (a molecule participating in a chemical reaction) for many biochemical and essential reactions and aids in the transport of materials and gases in the blood stream. In the blood stream, the amount of water pressure has a large impact on blood pressure and heart activity, both of which are monitored closely and can lead to changes in blood volume by your kidneys.

## Calcium Phosphates

Often ignored during the discussion of important molecules in the body, **calcium phosphates** make up much of the inorganic material in the bones and teeth. Certainly, the functions of these body parts are essential for support, movement, and eating, but it is important to note that storing calcium as phosphates is also a major bodily function of bones. Calcium is an essential ion for muscle contraction, nerve signaling, and protein activation, among other activities. If blood calcium levels decrease, calcium can be recruited from storage in the bones to maintain homeostasis of cellular activity. Glands secrete hormones that closely regulate blood calcium levels.

A) Glutamic acid

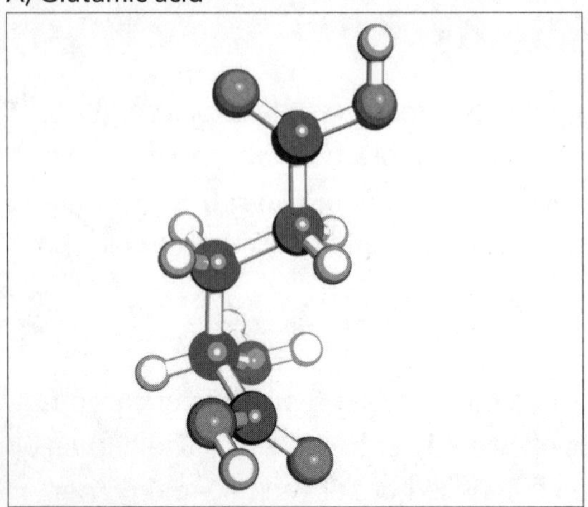

B) Cholesterol

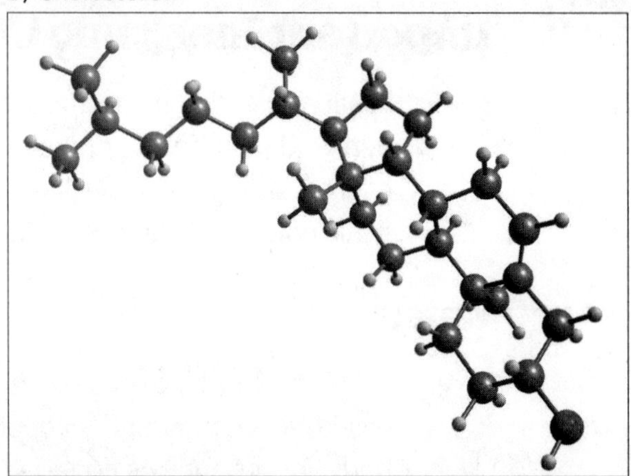

C) DNA

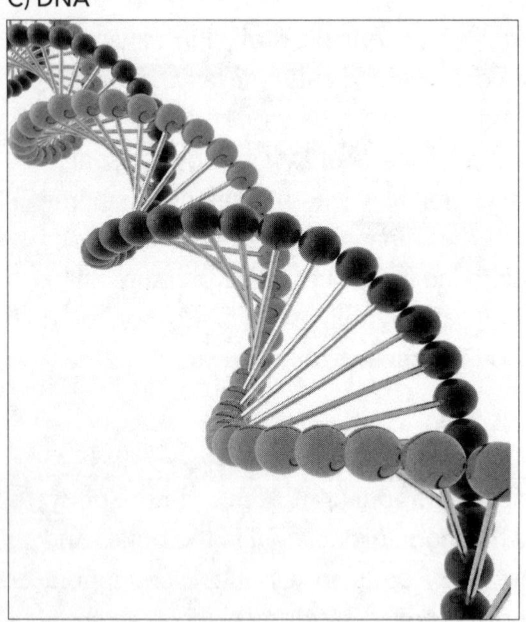

D) Glucose

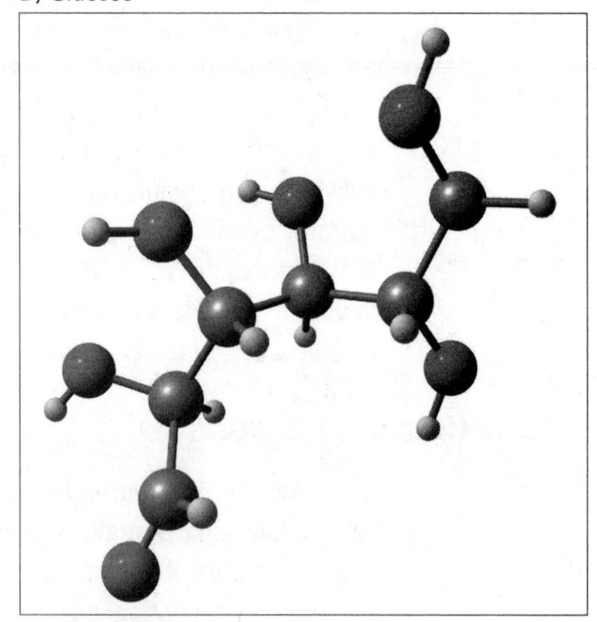

The four foundational organic molecules upon which all life is based are A) amino acids/proteins, like glutamic acid, B) lipids, which include cholesterol, C) nucleic acids, such as DNA, and D) carbohydrates, including glucose.

## Carbohydrates

Also known as "sugars," **carbohydrates** (or saccharides) play a major role in energy conservation, transport, transfer, and storage. Plants capture energy from the sunlight, which enables them to assemble carbon molecules into carbohydrates. Therefore, this energy from the sun is locked within the covalent chemical bonds that form these important molecules. Humans, consumers of these materials, will break down these complex molecules into individual $CO_2$ molecules and recover the energy from the bonds to be used elsewhere in the body. In your body, energy can be stored either as fat or as long chains of carbohydrates (**polysaccharides**).

A single carbohydrate molecule is often referred to as a **monosaccharide** with a typical chemical composition of $(CH_2O)_n$ where n is at least 3. Therefore, $C_3H_6O_3$, the simplest of monosaccharides, is called glyceraldehyde. **Glucose**, one of the most important energy-bearing monosaccharides, is $C_6H_{12}O_6$. These molecules occur and can be illustrated in either a ringed structure or a linear arrangement, which makes them nicely suited for storage and assembly into longer chains (polymerization).

**Disaccharides** are composed of 2 monosaccharides. Sucrose is composed of glucose and fructose, another important monosaccharide for metabolism.

The common name for sucrose is table sugar. It's granular in nature and used as a sweetener for food. Lactose is another common disaccharide found in milk that can cause digestive issues for those who are lactose intolerant.

While **oligosaccharides** are composed of between 3 and 9 monosaccharide units, polysaccharides may be much longer. In plants, cellulose and starch are the structural and storage forms of carbohydrates, respectively. For humans, other organic molecules are used to support the body. However, polysaccharides still provide energy storage for humans. Glucose is polymerized into glycogen, which is stored intracellularly (inside the cell) in muscle and liver cells to be broken down in times of high-energy expenditure and lower blood glucose levels.

# Proteins

Proteins are diverse and essential biological molecules for all cells. They serve as structural elements both inside (i.e., **cytoskeleton**) and outside of the cell (i.e., **extracellular matrix**): as anchoring molecules to hold cells in place, as adhesive molecules to allow cells to move from one place to another in the body, and as enzymes that facilitate much of the metabolic activity of the cell.

## Amino Acids

Amino acids are often classified into nonpolar, polar, acidic, or basic, depending on their R group. Carbon molecules are at the core of amino acids, which are the building blocks (**monomers**) of proteins. Amino acids are linked together via peptide bonds to form amino acid chains (**polymers**) called proteins. With 20 naturally occurring amino acids, the final shape and function of a protein is determined by the composition of the R group of the amino acids. Recall that of the 4 bonds that the central carbon forms at the core of an amino acid, the only one that differs between amino acids is that of the R group. The R group ranges from a single hydrogen atom in glycine to a ring of atoms in phenylalanine.

Several amino acids, such as valine and isoleucine, have **hydrocarbon** (i.e., molecules consisting of *only* carbon and hydrogen) **R groups**. These groups are neutrally charged and do not interact with charged or polar molecules, such as water. Therefore, these amino acids are said to be hydrophobic. They often interact with the cell membrane (which is also composed of hydrophobic hydrocarbon chains).

Other amino acids have an impact on the shape of the final structure that proteins are folded into. For instance, glycine has the smallest of the R groups, with only a single hydrogen present. This will allow the protein to fold into a more compact protein, since there isn't a large R group to physically get in the way.

| Amino acid | Abbreviations | | Molecular formula | Linear formula |
|---|---|---|---|---|
| Amino acid basic formula | | | $C_2H_4NO_2$-R | NH2-CH-R-COOH |
| Alanine | Ala | A | $C_3H_7NO_2$ | CH3-CH(NH2)-COOH |
| Arginine | Arg | R | $C_6H_{14}N_4O_2$ | HN=C(NH2)-NH-(CH2)3-CH(NH2)-COOH |
| Asparagine | Asn | N | $C_4H_8N_2O_3$ | H2N-CO-CH2-CH(NH2)-COOH |
| Aspartic acid | Asp | D | $C_4H_7NO_4$ | HOOC-CH2-CH(NH2)-COOH |
| Cysteine | Cys | C | $C_3H_7NO_2S$ | HS-CH2-CH(NH2)-COOH |
| Glutamic acid | Glu | E | $C_5H_9NO_4$ | HOOC-(CH2)2-CH(NH2)-COOH |
| Glutamine | Gln | Q | $C_5H_{10}N_2O_3$ | H2N-CO-(CH2)2-CH(NH2)-COOH |
| Glycine | Gly | G | $C_2H_5NO_2$ | NH2-CH2-COOH |
| Histidine | His | H | $C_6H_9N_3O_2$ | NH-CH=N-CH=C-CH2-CH(NH2)-COOH |
| Isoleucine | Ile | I | $C_6H_{13}NO_2$ | CH3-CH2-CH(CH3)-CH(NH2)-COOH |
| Leucine | Leu | L | $C_6H_{13}NO_2$ | (CH3)2-CH-CH2-CH(NH2)-COOH |
| Lysine | Lys | K | $C_6H_{14}N_2O_2$ | H2N-(CH2)4-CH(NH2)-COOH |
| Methionine | Met | M | $C_5H_{11}NO_2S$ | CH3-S-(CH2)2-CH(NH2)-COOH |
| Phenylalanine | Phe | F | $C_9H_{11}NO_2$ | Ph-CH2-CH(NH2)-COOH |
| Proline | Pro | P | $C_5H_9NO_2$ | NH-(CH2)3-CH-COOH |
| Serine | Ser | S | $C_3H_7NO_3$ | HO-CH2-CH(NH2)-COOH |
| Threonine | Thr | T | $C_4H_9NO_3$ | CH3-CH(OH)-CH(NH2)-COOH |
| Tryptophan | Trp | W | $C_{11}H_{12}N_2O_2$ | Ph-NH-CH=C-CH2-CH(NH2)-COOH |
| Tyrosine | Tyr | Y | $C_9H_{11}NO_3$ | HO-Ph-CH2-CH(NH2)-COOH |
| Valine | Val | V | $C_5H_{11}NO_2$ | (CH3)2-CH-CH(NH2)-COOH |

## Structure

Proteins have several levels of organization that begin with the amino acid sequence (which forms the **primary structure** of the protein). **Methionine** will always be the first amino acid in a protein chain since it is also the sequence (in the

RNA) that signals the start of protein formation. This sequence will have a profound effect on the remaining structure of the protein as well as its eventual function.

While the protein sequence is usually written in a straight line, proteins are flexible and typically either fold back upon themselves into side-by-side runs of the protein, which forms what is called a **beta pleated sheet**, or twist around neighboring regions of the protein into spiraling tubes, which are called **alpha helices**. Each of these folded patterns is held in place by hydrogen bonds between amino acids that are in close enough proximity for the hydrogen bonds to form. This arrangement of hydrogen bonds and particular amino acids forms the basis of what is termed the **secondary structure** of the protein.

In the same manner, other amino acids in distant regions of the primary sequence may become closer in proximity due to the folding of the protein. For example, when **cysteine** (which has a sulfate group) is next to another cysteine, it may form a disulfide bond (bonds that form between the sulfate groups of adjacent cysteine amino acids). When amino acids bond in this way, large loops of protein are held in place. This formation of larger folded structures within the protein is called the **tertiary structure** of the protein.

Lastly, separate protein units may be held together by bonds into large protein aggregates with each individual protein called a subunit. This bonding of individual protein subunits into a large protein aggregate is classified as the **quaternary structure** of protein and is illustrated in the hemoglobin molecule. Adult hemoglobin is formed from 4 subunit proteins, bonded together into the large single hemoglobin molecule. Its function is to transport oxygen in the blood stream.

# Lipids

Lipids are those hydrocarbon molecules that are used in plasma membranes and for energy storage in the human body. Because they are composed of hydrocarbon chains, much like those of some amino acids' side chains, they are hydrophobic in nature.

## Saturated versus Unsaturated Fatty Acids

Fatty acid chains are polymers of hydrocarbons, which are attached to a **carboxylic acid** (i.e., COOH). The carbons can be attached to each other via a single bond, with the remaining bonds completed with hydrogen molecules.

This would generate a saturated fatty acid in which all of the bonds of carbon are occupied by additional atoms. A straight linear fatty acid chain is generated when the carbon bonds are saturated. However, a double bond may be present between carbons (resulting in 1 fewer hydrogen atom on each of the adjacent double-bonded carbons). This is an unsaturated fatty acid and will have a bend at each of the double-bonded regions. Any more than 1 double bond in a single fatty acid chain will result in a polyunsaturated fatty acid.

## Phospholipids

Phospholipid is the main constituent in membranes, including the plasma, nuclear, mitochondrial, and vesicular membranes. Phospholipids are made when 2 parallel fatty acid chains are attached to a glycerol molecule. This not only organizes the fatty acid chains; it also provides a platform onto which a charged phosphate group is attached. Now, one end of the molecule is charged (i.e., hydrophilic) and the other end is composed of fatty acid chains (i.e., hydrophobic). This duality in the ability to interact with water is called being **amphipathic**.

## Triglycerides

Triglycerides are the storage form of energy in the body and typically are referred to as "fat." This material is stored in fat cells called **adipocytes** and can be recruited when your body requires the use of a lot of energy. Releasing double the amount of energy as compared to glucose, triglycerides take longer to release energy than carbohydrates do because the cells take longer to break down the fat and place it into the blood stream.

As the name implies, these molecules are composed of 3 fatty acid chains that are organized and attached to a glycerol molecule.

## Sterols

In humans, the principal sterol is cholesterol. While this substance is often crucified in the health world, your body cannot function without cholesterol. It plays a critical role in the proper spacing of the molecules in the plasma membrane, which will give more or less stability to the membrane under specific temperature ranges. Additionally, hormones such as estrogen and testosterone are derived from cholesterol, and are crucial to proper body function.

# Nucleic Acids

Nucleic acids are necessary for life and survival for each and every cell of the body. Nucleic acids exist in 2 forms, deoxyribonucleic acid (DNA) and ribonucleic acid (RNA). These linear molecules are the repository of genetic information (DNA) and copies of that information with which to build proteins (RNA).

## DNA

Described as a double helical molecule, DNA has 2 chains held together by hydrogen bonds that can easily be separated for either DNA replication (during cell division) or transcription (producing an RNA copy of the DNA). Each of the strands will be used as a template that the metabolic machinery of the cell can use to create new strands that result in an exact double helical copy of the parental DNA. Each set then will contain 1 original strand and a newly synthesized strand. These strands will end up in one of the 2 daughter cells that result from cell division.

DNA strands are composed of a few basic units. First, a sugar molecule will form part of the backbone of the strand of nucleic acid. For DNA, that sugar is deoxyribose, which represents the "D" in DNA. The other portion of the backbone is a phosphate group, which will link the sugars together into a longer and longer strand. Also attached to the sugar is a **nucleobase** of either a **purine** or a **pyrimidine**.

The purines of DNA are either adenine (A) or guanine (G), and the pyrimidines are thymine (T) and cytosine (C). The hydrogen bonds between the nucleobases are what hold the 2 DNA strands together into the double helix. The bases are always paired together into the purine-pyrimidine pairs A-T and G-C. Because of their structure, the A-T pair is held together with 2 hydrogen bonds and the G-C pair with 3 hydrogen bonds. Therefore, the G-C pair will require more energy to break than the A-T set. This becomes important for DNA replication and RNA synthesis.

## RNA

RNA is similar in its structure to DNA with some important differences. As the name implies, the first difference is the sugar used in the backbone. For RNA, that sugar is ribose. Additionally, RNA will be synthesized as a single strand rather than a double strand. Lastly, while G, C, and A are found in RNA, T will not be present. Rather, another pyrimidine is used instead of T—uracil (U).

CHAPTER 2

# Cells

All living things are composed of one or more cells. Since they are the basic unit of life, cells allow for diversification of shape, function, and activity. Therefore, multicellular organisms can be so extremely complex and unique. In the following sections we explore all of the essential component parts of cells—their molecular constituents and their organization—and how these organelles function collectively to keep the cell alive. Additionally, this chapter examines how cells traffic materials into and out of the cell during normal metabolism. Finally, the chapter will examine how cells replicate themselves during a normal life cycle as well as the production of cells used in sexual reproduction.

## Components of a Cell

A **eukaryotic cell** is a cell with a membrane-bound nucleus. The components of this type of cell are necessary for the segregation and diversification of cellular function and activity. For instance, it would be counterproductive to have degradative enzymes (which could destroy cell material) mixing with important genetic material. Therefore, cells separate these biological activities within defined compartments so that multiple reactions can occur in a defined space without interfering with each other. Cells consist of several **organelles**, which are physically and functionally distinct compartments of the cell, including, but not limited to, the membrane, cytoplasm, nucleus, endomembrane, and mitochondria.

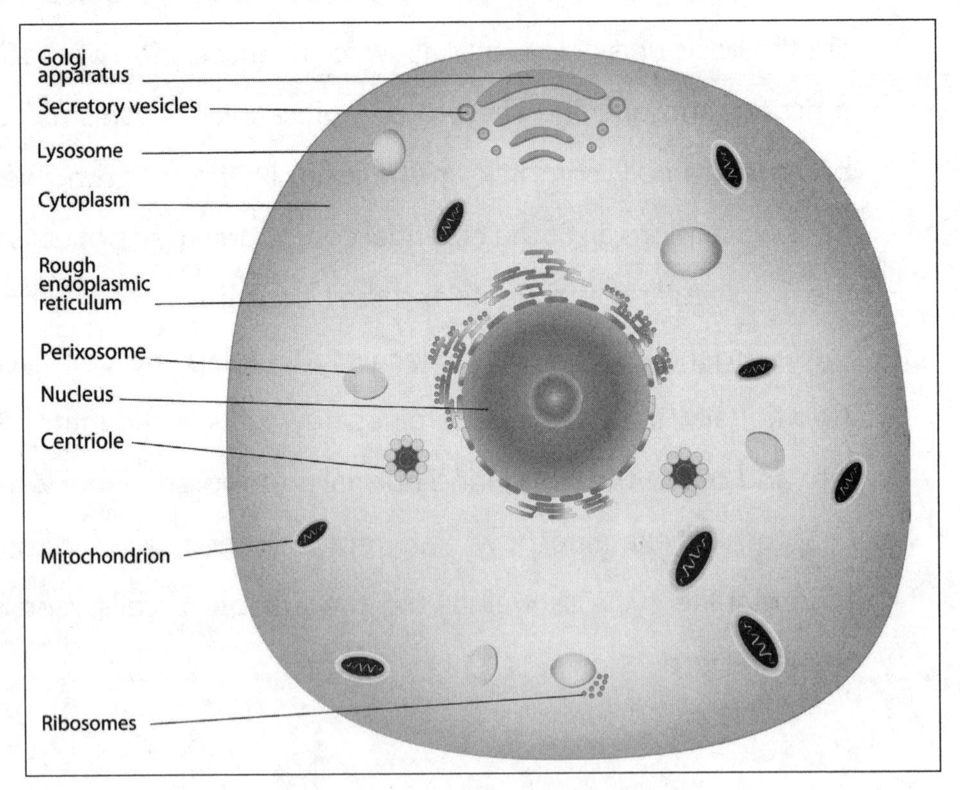


A typical animal cell with organelles.

## Membranes

The **cell membrane** or **plasma membrane** forms the boundary between the inside and outside of a cell. It consists of both protein and lipid molecules in varying ratios depending on the cell type and species. Typically, there are 50 lipid molecules for each protein in the membrane. However, since proteins are much bigger than lipids, proteins make up 50 percent of the mass of the membrane. These molecules are arranged in 2 opposing sheets, creating a bilayer. One layer faces the outside of the cell, or the **extracellular** surface, and the other faces the interior, the **cytosolic** surface.

The main lipid type in the membrane is a **phospholipid**, a molecule made of a charged phosphate group attached to a base foundational molecule (either a glycerol or sphingosine). This is the portion of the lipid bilayer that faces the outside of the cell for one sheet and the inside of the cell for the opposite sheet. Because of the negative charge of the phosphate, this face of the phospholipid layer can interact with water molecules, which are also charged, or **polar**). With the charged groups of the lipid layers facing outward and inward, the fatty acid chains of each lipid layer becomes sandwiched between this charged surface and create a nonpolar (**hydrophobic**) layer. Just as oil and water don't mix, this layer repels water or charged molecules and makes an effective filtration barrier, called a **semipermeable membrane**. The charged surfaces, both outside and inside, allow for the interaction of polar and charged molecules with the membrane, while the hydrophobic internal portion of the bilayer creates a semipermeable membrane through which only certain molecules (i.e., oxygen, carbon dioxide, and water) can freely diffuse.

Cholesterol is abundant in the plasma membrane and serves to regulate membrane fluidity. The shape of the hydrocarbons in cholesterol prevent the phospholipids from packing together too tightly, which would result in a more rigid membrane.

Proteins embedded in the phospholipid bilayer may be associated with either or both of the surfaces of the membrane. This membrane-spanning arrangement enables the proteins to serve many cellular functions. They transport material into or out of the cell and provide membrane attachment for stationary cells or adhesion for migratory cells. Many proteins in the membrane are receptors that recognize chemical signals and relay those signals to the inside of the cell to alter cellular activity.

Proteins in the membrane are not locked in place. They may float throughout the membrane, spin, or flip horizontally. The fact that proteins may move freely throughout the membrane is called the **fluid mosaic model**.

## Cytoplasm

Effectively separated from the extracellular environment by the plasma membrane, the inside of the cell is the location of most metabolic activity. The cytoplasm houses the workshops of the cell. Here, incorporated material is broken down, new proteins are generated, and new phospholipids are produced. All of these activities are compartmentalized via membranes within the cytoplasm and will be explained in the following section.

## Nucleus

The membrane-defined **nucleus** is positioned in the center of the cell. The nucleus contains the **DNA**, which is the genetic code for the cell. Responsible for the protection and replication of this material, the nucleus produces copies of the DNA code into **RNA**, which is used in the cytoplasm to make proteins.

The membrane surrounding the nucleus consists of the same materials as the plasma membrane. But the nuclear membrane is composed of 4 phospholipid layers (2 bilayers) with a **perinuclear** space between. The presence of these 2 bilayers led to the endosymbiosis hypothesis. According to this hypothesis, during evolution, a cell lacking a membrane-bound nucleus (a **prokaryotic** cell) engulfed a cell possessing and capable of managing a large amount of DNA. As the prokaryotic cell engulfed the second, it wrapped the additional bilayer around its new partner (the nucleus). As a result, the 2 cells formed a symbiotic relationship in which both benefited equally from the shared resources and products they recieve from each other.

When you look at a nucleus using a light microscope, you might see a dark spot resembling the pupil of an eyeball looking back at you. This most prominent structure within the nucleus is called the **nucleolus** and is made up of stretches of densely packed genetic material and proteins. The density of the nucleolus, which accounts for its dark staining, is produced during synthesis of ribosomal subunits.

Just as protein channels within the plasma membrane facilitate transport of materials into and out of the cell, protein complexes within the nuclear membrane regulate transport into and out of the nucleus. While small water-soluble molecules may pass unimpeded, larger molecules must be assisted to get from one compartment to the other. The "helper" is a **co-transport** molecule, which must bind to the "cargo" molecule to allow the transport. The helper that moves molecules into the nucleus is called **importin**, and **exportin** moves molecules out of the nucleus.

## Endomembrane System

Many of the membrane-bound organelles are either physically or functionally connected or both, and so are grouped together as part of the endomembrane system. The nucleus is often considered to be part of this system because of its physical connection with the endoplasmic reticulum. The remaining members of the endomembrane system are the ER, Golgi, and vesicles.

### Endoplasmic Reticulum

The **endoplasmic reticulum (ER)** functions in protein and lipid production. It is made up of large, folded sheets of membranes that occupy vast expanses of the cytoplasmic compartment. The folds of the ER create a space between the membranes that is continuous with the perinuclear space (the space between the 2 bilayers of the nuclear membrane). Therefore, the membrane of the ER is continuous with the outer nuclear membrane. To illustrate this concept, think of the ER membrane as extensions of the outer nuclear membrane projected into the cytoplasm. Also, the internal portion of the ER, as defined by its own membrane, is continuous with the space between the inner and outer nucleus membrane.

There are 2 types of ER: rough ER (**rER)** and smooth ER (**sER**). Rough ER is covered with **ribosomes**, the organelles for protein synthesis, which give the ER a rough appearance. Material synthesized within the rER then must make their way to the next component of the endomembrane system, the **Golgi apparatus**. Membranes without ribosomes are called smooth ER and are the sites of lipid synthesis. However, since these organelles are physically separated within the cytoplasm, material of the ER is transported in membrane-bound spheres called vesicles that form from the rER membrane and will move toward and fuse with the membranes of the Golgi apparatus. In other words, portions of the ER will pinch off as small membrane-bound packages called vesicles that are filled with protein that has been synthesized in the ER. These vesicles will diffuse to the membrane of the Golgi apparatus, fuse together to create a new stack of Golgi membrane, and continue the process of protein modification in this organelle. In this way, proteins and lipids are generated. They supply the cell with fresh resources with which to continue to survive.

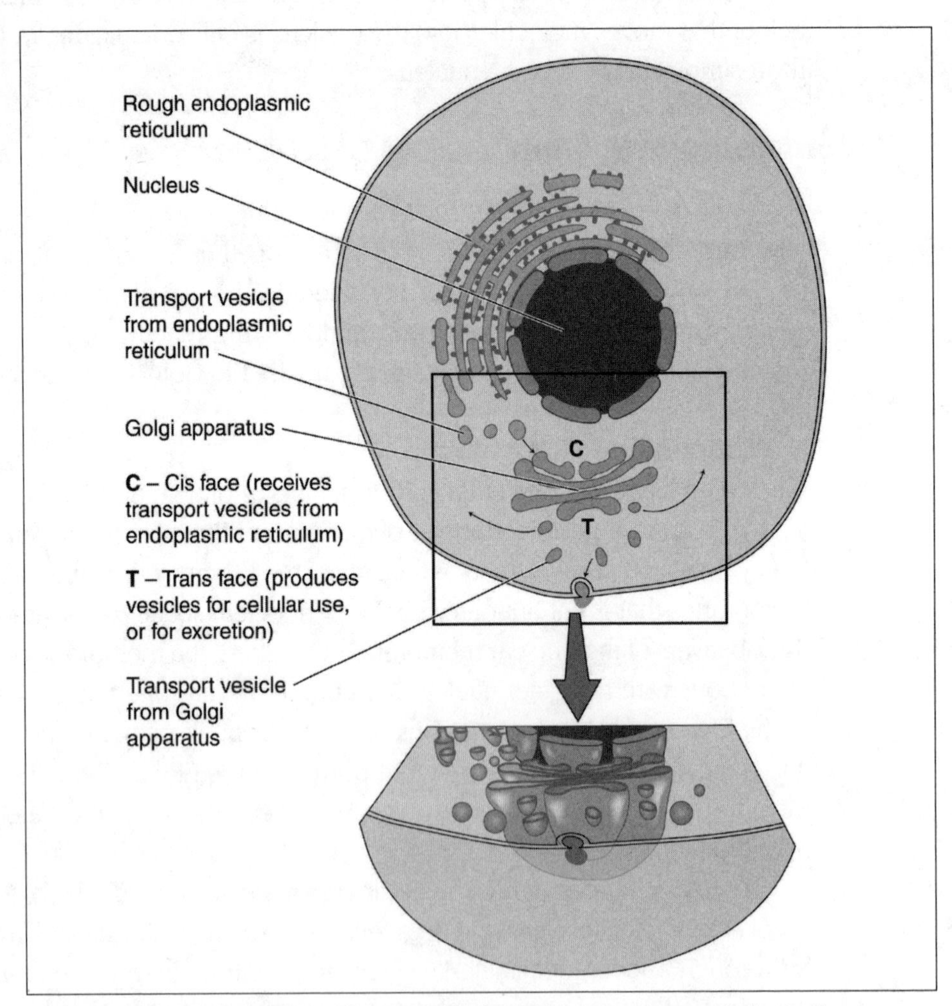


Protein synthesis can begin on the ER (folded membranes connected physically to the nuclear membrane). Transport vesicles shuttle material from the ER to the Golgi apparatus to continue the process of protein modification. Finally, vesicles from the Golgi ship the newly finished materials to the proper compartment in the cell.

## Golgi Apparatus

Proteins of the ER are transported in membrane-bound spheres called vesicles that form from the rER membrane and will move toward and fuse with the membranes of the Golgi apparatus.

Shaped like a stack of pancakes, the Golgi is formed by the aggregation of transport vesicles from the rER into the stacked structure typical for the Golgi apparatus. As the vesicles fuse together on the incoming side, they form a new layer termed the **cis face** of the Golgi. Like an assembly line, these layers are moved higher and higher in the stacks of the Golgi as new cis layers are added. At the opposite side of the stacks, the **trans face**, or last layer, breaks up into transport vesicles and shuttles material to its target.

As proteins are passed along the layers of the Golgi, proteins continue to be modified, a process that began in the rER. These modifications include **glycosylation** (the addition of carbohydrates) and/or the addition of phosphates (**phosphorylation**). The Golgi is also essential for sorting proteins and packaging them for specific targets. Some materials will be shipped to the plasma membrane, while others will be placed into vesicles going to other membrane-bound organelles, such as the mitochondria.

## Vesicles

In addition to the transport vesicles of the endomembrane system, other vesicles are essential for proper cellular function. **Lysosomes** are spheres of enzymes that break down proteins, carbohydrates, or fats. These enzymes are emptied into other vesicles containing engulfed materials or into membrane-bound organelles (e.g., mitochondria) that are signaled for destruction and recycling.

**Peroxisomes** are another group of vesicles that contain hydrogen peroxide. These are present in both plant and animal cells and are prominent in liver and kidney cells, where the hydrogen peroxide detoxifies ethanol and breaks down fatty acids.

## Mitochondria

The **mitochondria** are often called the powerhouse of the cell. Consisting of a double membrane system, much like the nuclear membrane, mitochondria are also thought to have arisen from a symbiotic relationship between 2 cells in the evolutionary past. Support of this hypothesis stems from the composition of the membranes of the mitochondria. While the outer membrane has a protein-lipid ratio similar to that of other animal cells, the inner mitochondrial membrane has a ratio similar to bacterial membranes.

Like the nucleus, mitochondria also possess DNA. The mitochondrial genome encodes for over 30 genes whose products play essential roles in metabolism and energy production.

Shaped like a capsule, the outer mitochondrial membrane is flat over the surface of the organelle while the inner membrane is folded (to increase surface area) into sheets, which are called **cristae**. Proteins in the inner membrane create an electron transport system where protons, or hydrogen ions ($H^+$), are transported from the interior of the mitochondria, called the **matrix**, to the **intermembrane space** (the space between the inner and outer membranes). Continued transport of $H^+$ will cause a pooling of the ions, and this pooling creates a gradient of protons that will flow through another protein complex of the inner membrane, the **ATP synthase complex**. The energy of the flowing $H^+$ is used to produce **ATP** (**adenosine triphosphate**), the molecule that all cells of the human body use for energy.

## Molecular Trafficking

From the transportation of nutrients and oxygen into the cell to the elimination of hormones and waste materials out of the cell, the plasma membrane is a site of dynamic and constant movement between the outside and inside of the cell.

## Diffusion

Simple **diffusion**, molecules moving down a concentration gradient without the use of energy, is much like riding a bicycle downhill. The only energy required was what was used to get up to the top of the hill. Afterward, it's simply letting gravity "coast" you downhill. For molecules, the push is the concentration of molecules. The top of the hill is therefore the area of high concentration and the bottom of the hill is low concentration. For instance, oxygen and carbon dioxide freely diffuse through the membrane during **respiration**. Also, water is capable of freely diffusing through the plasma membrane. The diffusion of water can also be called **osmosis**.

Water molecules tend to dilute materials to an equal extent. If water on one side of a membrane has more **solute** (a substance other than water such as protein, carbohydrate, ions, etc.) added, water will flow from the side of less solute (the **hypotonic** side) and into the side rich in solutes (the **hypertonic** side) in order to attempt to equalize the concentration of water and solutes on both sides of the membrane. This is a key mechanism used by the kidneys to conserve water in the body during urine production.

## Carrier Mediated Transport

Molecules that can't diffuse through the plasma membrane (such as glucose because it's too large) are transported through the membrane via **protein channels**. Also, many molecules must be transported upstream (i.e., against their concentration gradient), which also requires transmembrane proteins in addition to energy. Think of this like salmon swimming upstream to spawn. They expend a lot of energy to accomplish the task of reproduction. Cells move materials against the gradient and use energy for their own survival.

Glucose and charged ions such as sodium are among the molecules and ions that must use a protein channel for diffusion into or out of a cell. Since this is still diffusion, no energy is used; the only difference is the specialized tunnel through which these molecules can diffuse.

**Active transport** is a type of transport that is distinct from diffusion in that molecules are transported against a concentration gradient. This means that molecules must travel from an area of low concentration to a high-concentration area, which requires energy. Active transport protein channels, which can be thought of as tunnels or hollow pumps, bind the transport molecules and use energy from ATP to change the molecules' shape in such a way as to move them across the membrane and against the gradient. The final step of the active transport process is to reset the pump so that the next active transport cycle may begin.

The sodium-potassium pump facilitates the active transportation of 3 sodium ions out of the cell and moves 2 potassium ions into the cell during the resetting of the pump. The sodium-potassium pump accounts for approximately 12 percent of the energy utilization of the entire human body, as it is involved in neuronal signaling, muscle contraction, and kidney function.

## Cell Growth and Replication

Cellular growth and division is a simple fact of life itself. Controlled at the molecular level and via secreted materials, growth and division are maintained with precision, unless a cell or group of cells begins to divide out of control, from a genetic mutation, which leads to the formation of a tumor.

### Cell Cycle

A typical cell will spend the majority of its cell cycle providing an essential function to the tissues and organs where it resides. The first stage of the cell cycle is called **interphase**: the growth and synthesis portion of the life of a cell. In this stage, the cell grows to its final size and may remain in this static, functional state until it prepares to divide again. Following the last portion of interphase, cells then divide their chromosomes and nucleus. This is the **mitosis** stage. The cell cycle finishes with **cytokinesis**, the division of the cytoplasm, resulting in 2 daughter cells identical to the single parent cell from which they were produced.

### Interphase: G1

Both mitosis and interphase (which are stages of the cell cycle) can be broken down further into phases to help in the understanding of these complex activities. Once mitosis and cytokinesis are complete, the daughter cells enter the first phase of interphase: the **gap 1 phase**, or **G1**. During this important period, most cells increase in size, replicate essential organelles such as the mitochondria, and move the nucleus more toward the center of the cell. Late in this phase, cells are screened for size and any DNA damage by regulatory molecules of the cell before being allowed to move to the next portion of interphase.

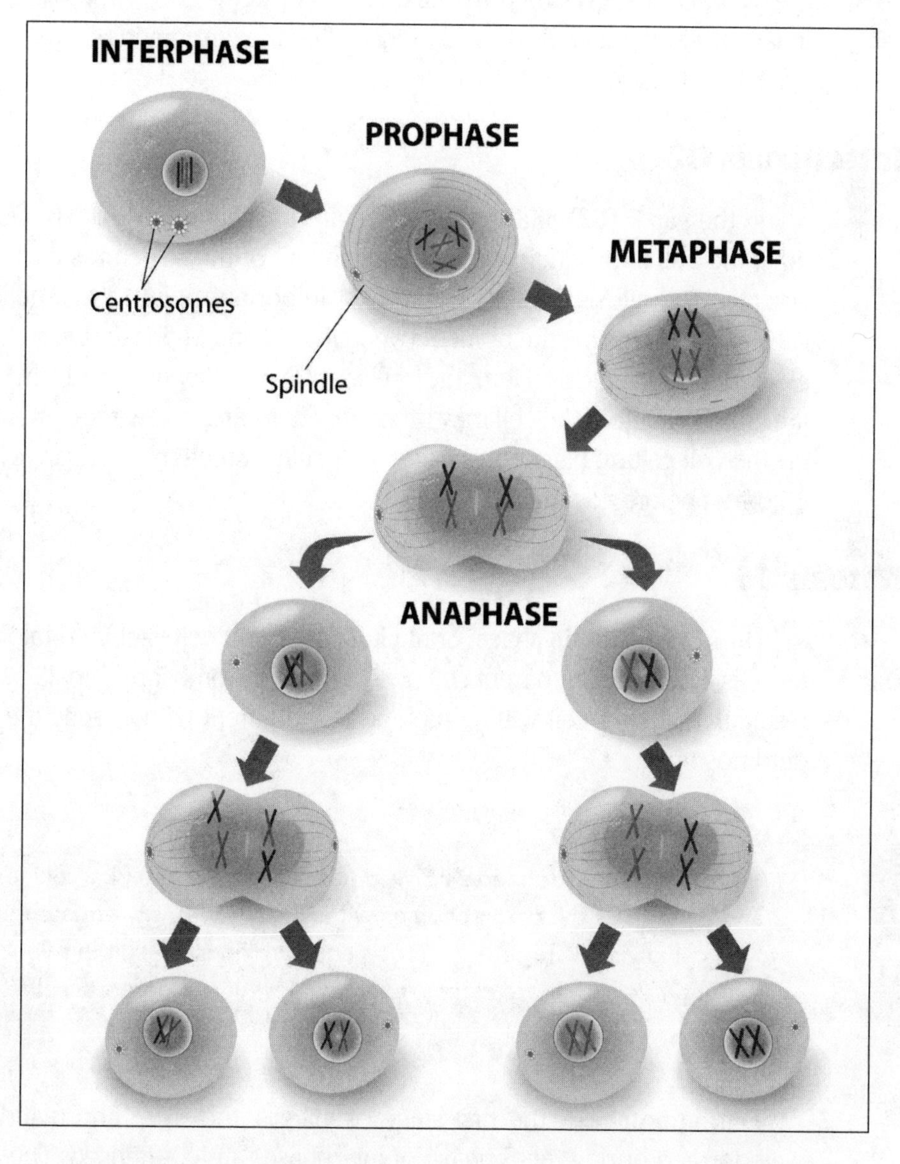


A cell undergoing mitosis, forming 2 identical daughter cells. The daughter cells undergo mitosis, creating 4 cells that are identical to the original parent cell.

## Interphase: S

The **synthesis phase**, or **S phase**, follows G1 and is the period where the chromosomes are copied to enable each daughter cell to have a full complement of DNA. Each linear (straight) chromosome is copied and remains attached at the central point of the chromosome, or **centromere**, resulting in the typical X-shaped structure that many incorrectly term chromosomes when, in fact, they are sister chromatids. Only when the chromatids separate are they again chromosomes.

## Interphase: G2

In the **gap 2 (G2) phase**, the cell begins to prepare for mitosis. During this time, the cell is producing and organizing all of the structures and materials essential for mitosis. However, the most important point in this phase is the G2-M transition point, the point between the G2 and M phases before the cell is cleared to start mitosis. Here, the cell size, DNA replication, and DNA damage are checked before the cell may proceed. These aren't new processes, but simply the cell getting bigger and using molecular screeners to ensure everything is correct before advancing.

## Mitosis: M

During the **mitotic phase**, or **M phase**, a parent cell is cloned into 2 daughter cells. Each human parent cell possesses 46 chromosomes in 23 pairs. Each resulting daughter cell will be a clone of the parent with exactly the same 46 chromosomes.

Mitosis is strictly defined as division of the nucleus (i.e., karyokinesis) and division of the chromosomes within. Many wrongly equate mitosis with cell division. Together, mitosis and cytokinesis result in two daughter cells that are clones, exact copies, of the parental cell.

During **prophase**, the first stage of mitosis, the sister chromatids (which were formed during the S phase of interphase) are condensed. The DNA will continue to compact throughout this period and then next into the visible

X-shaped sister chromatids. Additional changes occur in the cell during this phase. As the nuclear membrane begins to disintegrate, fibers in the cytoplasm that will be used to pull the chromatids apart begin to assemble.

As somewhat of an intermediate period between 2 stages, **prometaphase** is often considered late prophase by many scientists. During this time, centrioles begin to organize microtubules that stretch from each pole of the cell toward the opposite pole. These are the **spindle fibers**. The spindle fibers will serve 2 critical functions during cell division. First, fibers from each pole will connect to each side of the sister chromatids at the centromere to move the chromatids to the middle of the cell and then pull them apart during a later anaphase stage. Other spindle fibers will interact with fibers from the opposing pole of the cell and function to push the poles of the cell farther apart in preparation for the division of the cytoplasm in the last stage of mitosis.

The most often illustrated phase of mitosis, **metaphase**, the middle phase of mitosis, is recognizable because of the alignment of all 46 sister chromatids at the equator of the cell. These are clearly visible in the middle of the cell as they pause in preparation for separation of the chromatids in the next stage, **anaphase**.

The shortest of the phases of mitosis, anaphase, is characterized by the pulling apart of the sister chromatids into 46 individual and identical chromosomes that are being moved in opposite directions. Movement of the chromosomes is relatively rapid as the spindle fibers pull the chromosomes toward the poles at approximately 1 mm/min (micrometers).

Anaphase continues until the chromosomes arrive at the poles, which signals the start of **telophase**. At this time, the chromosomes begin to rapidly relax and uncoil, the nuclear membrane begins to re-form, and many of the spindle fibers disappear. This concludes the division of the nucleus, which coincides with the initiation of cytoplasmic division, or cytokinesis.

## Cytokinesis

In animal cells, near the end of telophase, cytoskeletal proteins called **actins** begin to form into a belt that extends around the equator of the cell. As cytokinesis continues, the actin ring begins to get smaller and smaller, resulting in a narrowing at the waist of the cell called the cleavage furrow. This constriction continues until the 2 resulting daughter cells are pinched away from each other into independent and free identical cells.

# Meiosis

For the continuation of the species and during sexual reproduction, the genome (total genetic content) of the **germ cells** (sperm and egg) must be divided in half. This creates 2 **haploid cells**, containing one copy of each chromosome, so the resulting new individual will have the final complete amount of genetic material. For humans, that is 2 copies of each chromosome, 1 from the mother and 1 from the father. Therefore, while mitosis is often referred to as cloning, meiosis is termed a reduction division.

This reduction of genomic material is accomplished by undergoing 2 divisions, each named as in mitosis followed by the division number (e.g., prophase I, metaphase I). The first division separates **homologous chromosomes**. These are the chromosomes that occur in pairs, 1 from the mother and 1 from the father. For example, during meiosis I, the homologous chromosome 1 from the mother will be pulled into one new cell while the homologous chromosome 1 from the father will be moved into the opposite cell. This will occur independently for each of the 23 pairs of homologs in the parent cell.

The second division is identical to mitosis in that each daughter cell from the first division will now have its sister chromatids pulled apart. The only difference between this division and mitosis is that the starting material for meiosis II are 23 sister chromatids (in mitosis there are 46). Therefore, after the completion of 2 divisions, 4 haploid (1 chromosome each) daughter cells will result.

## Meiosis I

During prophase I of meiosis, the arms of homologous chromosomes may become entangled and form a cross-shaped tangle point (**chiasmata**), resulting in the exchange of large portions of homologous chromosomes in a process called **crossing over**. This is a normal process that results in increased diversity of a species. However, if problems do happen, such as a crossing over between the wrong chromosomes, genetic defects can occur.

Prometaphase I in meiosis is the same as in mitosis. However, while the sister chromosomes align at the equator during metaphase I, their arrangement is slightly different than in mitosis. In this stage, 23 pairs of homologous chromosomes are stacked at the equator rather than all aligned in single file as in mitosis. In anaphase I, the homologs are being pulled toward the opposing poles of the cell. Telophase I and cytokinesis are identical to mitosis with the only exception being that the nuclear membrane does not re-form.

Homologous chromosomes, homologs, and homologous pairs are terms that all refer to those chromosomes composed of the same genes. For example, chromosomes number 1 from the mother and number 1 from the father represent one of the homologous chromosome pairs of the human genome.

## Meiosis II

This second division begins with 2 daughter cells, each comprising 23 sister chromatids, which will be separated in the same stages and in the same manner as was observed in mitosis. Thus mitosis and meiosis II only differ in the number of sister chromatids that must be pulled apart. Following prophase II, metaphase II, anaphase II, and telophase II (including cytokinesis), the 2 daughter cells will have divided into a total of 4 haploid daughter cells.

CHAPTER 3

# Replication and Metabolism

The biochemical processes that cells use to survive, replicate themselves, and produce energy are highly complex and co-ordinated events. This chapter will simplify these mechanisms and highlight some of the major hurdles that stump many students. Additionally, you'll find some hints and tips that will help you to confidently learn and remember the steps of the processes.

# DNA Replication

Before a cell can divide into 2 daughter cells, all of the DNA of the parent cell must be copied so that each daughter gets a copy. Understanding the following terms will help you understand this process even more.

## DNA Structure

**DNA** (deoxyribonucleic acid) is a **double helix** (two spirals braided together) consisting of antiparallel (i.e., running in opposite directions) strands of nucleic acid. It might be helpful to picture the DNA molecule as a ladder. Each leg of the ladder represents one strand of DNA and the rungs represent the nucleotide bases that hold the entire structure together. In DNA, the nucleotides on one strand are always paired with a partner in a very consistent manner. Since DNA pairs contain only 4 different nucleotides, they always exist as adenine (A): thymine (T) pairs or guanine (G): cytosine (C) pairs.

While a ladder is usually viewed as either a way up or a way down, in the DNA analogy in the previous paragraph, each leg of the ladder will be leading in the opposite direction. One leg will be pointing up while the other leg is pointing down. When climbing a straight ladder, which leg is on the left and which is on the right is irrelevant, but for DNA and its replication, it's essential for determining the direction new DNA strands will be produced.

DNA strands exist as a "one way" street that runs 5´ to 3´ in biochemical terms. So, while one strand (the coding strand) is read left to right as 5´ to 3´, the other strand of the double helix (the complementary strand) is present 5´ to 3´ in the opposite direction. The 5 and the 3 are designated carbons of the free end of the molecule and thus determine the direction with which the DNA can be assembled.

## Preparing for Replication

Before the DNA double helix can be replicated, it must be separated into single strands. This is likened to unzipping a pair of pants. On the DNA molecule, the enzyme DNA **helicase** will bind and separate the nucleotide pairs, freeing one strand from its partner. Although both strands are being replicated at the same time, for simplicity the strands and their replication will be considered separately.

## Leading Strand Replication

First let's talk about the complementary strand in regard to DNA replication. This strand is the "opposite" partner for the genetic code (i.e., the coding strand). But, it will be used to as a template to produce a new double helix for one daughter cell.

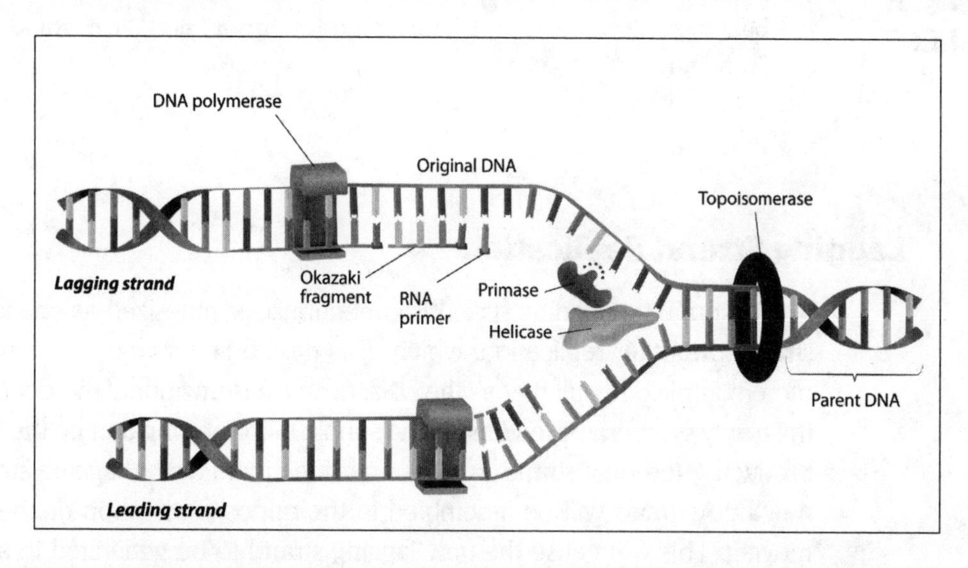


DNA replication is illustrated at the enzymatic process by which a DNA molecule (composed of two strands) is copied into two double-stranded DNA molecules identical to the parental DNA.

By using both strands as templates, each daughter cell will contain one new strand and one inherited strand from the parent, thereby forming the basis of semi-replicative division.

With the DNA unzipped, an enzyme called DNA polymerase (pol) binds to and reads the template strand in a 3´ to 5´ direction, and assembles a new coding strand in a 5´ to 3´ direction. For example, if the sequence on the complementary DNA strand were 3´–A-T-C-G-G-T-T-A–5´, then the new coding strand would be assembled in this order: 5´–T-A-G-C-C-A-A-T-3´. Since this is also the direction in which the DNA helicase enzyme is unzipping the double helix,

DNA pol simply follows along until all of the strand has been copied into a long new coding strand, which is called the **leading strand**.

Nucleotides are always assembled in a 5′ to 3′ direction and the templates are read in the opposite direction. To remember what the numbers mean, just recall that 5 can stand for "front" and 3 can stand for "the end."

## Lagging Strand Replication

Although the leading strand is continuously synthesized as one long new strand, the other replication, which uses the coding strand as a template, is more complex. Recall that as the DNA helicase is unzipping the double helix, the newly synthesized leading strand simply follows along behind the helicase, producing the new strand in the same direction. For the lagging strand, the new DNA strand will be assembled in the opposite direction the helicase is moving. This will cause the new lagging strand to be generated in a discontinuous manner. For example, as soon as an area of single-stranded DNA is exposed, a new strand is being generated. But in this case, the new strand is being formed in the opposite direction from the way the helicase is unzipping DNA. Therefore, as the helicase unzips away from the new strand segments, called **Okazaki fragments**, a gap of single-stranded DNA will be present between the new strand and the helicase. This will lead to the initiation and generation of another Okazaki fragment, which will form and extend until the previous fragment is reached. This process will continue producing distinct Okazaki fragments along the length of this template into a discontinuous lagging strand with gaps remaining between the fragments. The **gluing**, or ligation, of these fragments into one long continuous strand is the function first of DNA polymerase and then of DNA ligase, which binds to the gap site between adjacent Okazaki fragments and ligates them all together into a complete strand.

# Transcription

The genetic code of a cell is housed in the nucleus as DNA. However, before this code can be read, interpreted, and used to create proteins, a copy must be made and shipped into the cytoplasm where protein synthesis occurs. The process of making this copy of the genetic code is called **transcription** and is very similar to DNA replication, with only a few differences.

During transcription, nucleotides are assembled into an RNA molecule by the enzyme RNA polymerase. This enzyme, which is much larger than DNA polymerase, is capable of binding to specific sequences of DNA, unwinding the DNA, reading a single strand of DNA as a template (i.e., complementary strand), and generating a single strand of RNA that will contain the genetic code for making protein. Once the RNA molecule is produced, it detaches from the RNA polymerase, and the DNA strands bind back to one another to form the double-stranded molecule.

While the nucleotides in DNA are A:T and G:C pairs, the nucleotide uracil (U) replaces thymine (T) in RNA. Thus, the nucleotide pairs in RNA are A:U and G:C.

The idea of a transcript in written language is to produce a word-for-word copy that uses the exact same language as is spoken. In the case of the genetic code and nucleotide alphabet, a copy in the form of RNA is the transcript, consisting of the same code as found in the coding strand of DNA.

Several types of RNA will be produced by transcription. Messenger RNA (mRNA) is the genetic code in RNA form. Ribosomal RNA (rRNA) will be combined with proteins to form ribosomes, the organelles for protein synthesis. Lastly, transfer RNA (tRNA) are molecules that transport amino acids in the correct order to the ribosome for assembly into protein.

# Translation

When a translation of a spoken language occurs, the communication happens in a completely different language, often using a different alphabet. The same is true in cellular translation. Here, the machinery of the cell reads the genetic language of nucleotides (of the mRNA) and assembles proteins using an alphabet of amino acids. The essential components of translation are the 3 RNA molecules: mRNA, rRNA, and tRNA. Translation is the essential, almost continuous, process of producing new proteins cells depend on for survival.

## RNA

Much like words are composed of letters, the genetic alphabet of nucleotides is arranged on mRNA into 3-nucleotide "words" called **codons**. With 4 different nucleotides arranged into 3-letter codons, there are 64 distinct codons that can be used by the translation machinery of the cell. While mRNA represents the code, the location where this code is read and interpreted and proteins are assembled is the **ribosome**.

Ribosomes are organelles in the cytoplasm of the cell that consist of 2 (1 large and 1 small) rRNA subunits along with associated proteins. Much like cupping 2 hands together creates a space between your palms, when the 2 rRNA subunits unite, they create spaces on the interior that will be used to hold onto the mRNA, as well as to import and assemble amino acids into proteins. For this reason, the ribosome is often considered the factory for protein synthesis. But, in the same way that a factory would produce nothing without workers, ribosomes would never synthesize proteins without tRNA. The tRNA brings the building blocks of proteins into the ribosome in the correct order as specified by the code of the mRNA.

tRNA molecules are able to bind to specific amino acids on one end while recognizing codons of the mRNA on the opposite end. Just as the coding strand of DNA binds to its complementary strand via nucleotide pairs (i.e., A:T, C:G), tRNA has 3 nucleotides (the **anticodon**) arranged that are complementary to the mRNA codons. This means that as the ribosome slides along the mRNA molecule, tRNA molecules bind to their respective codons, and in doing so, bring amino acids into the interior of the ribosome in the order directed by the mRNA code and, therefore, into the correct sequence for the protein.

The space inside the ribosome is only capable of holding 2 tRNA molecules at a time; so, as the ribosome slides along the mRNA, the amino acids become linked together (via **peptide bonds**), thus freeing one amino acid from its respective tRNA molecule and forming a growing chain of amino acids (i.e., protein) attached to the other tRNA. At this point, the empty tRNA molecule is ejected, the other tRNA and its growing chain slide into the adjacent space (vacated by the ejected tRNA), and a new tRNA molecule with an amino acid enters. This assembly will continue until the entire protein is finished.

**Given an mRNA codon of AUG, what would be the complementary anticodon of the tRNA molecule that would bind here?**
The anticodon that would recognize AUG would be UAC. Remember from Chapter 2 that there are no T nucleotides in RNA.

## Codons

Although there are 64 possible codons, only 20 amino acids can occur in nature and be assembled into proteins. Does this mean many of the codons are irrelevant? The answer is no. Multiple codons can be recognized by the same tRNA. In addition, multiple codons encode for the same amino acid. For instance, the codons GGU, GGC, GGA, and GGG are all recognized by the tRNA for the amino acid glycine. In fact, many tRNA molecules can recognize 4 different codons.

For simplicity, these are often illustrated and arranged into a codon table, also called the genetic code table.

Notice that for glycine, the 4 codons only differ in the last nucleotide. This flexibility is conferred by a fifth nucleotide that can be placed into the anticodon of tRNA: inosine. This nucleotide enables the "wobble effect," so that it can bind with any of the other 4 nucleotides of mRNA. Simply, several codons can encode for the same amino acid.

In addition to the codons for amino acids, there are also start and stop codons that signal the beginning and end of protein synthesis. The codon AUG is the one and only start codon, and signals the assembly of the first amino acid in all proteins: methionine. To stop translation, when any of the codons UAA, UAG, or UGA is read, it signals the disassembly of the ribosome complex, thus freeing the newly generated protein.

## Enzymes

**Enzymes** are a class of proteins made via translation. These molecules catalyze naturally occurring chemical reactions by making it easier to cut, modify, process, or further manipulate material into a final product.

### Activation Energy

Activation energy can be considered the threshold a chemical reaction must overcome in order to form a product. The enzyme lowers this threshold, and therefore less energy is required to create the product, the rate of the reaction increases, and the entire process becomes more efficient. In other words, enzymes save energy and make the metabolic job of a cell much easier and faster.

No enzyme can catalyze a chemical reaction that would not occur in nature. Enzymes only enable the reaction to occur faster.

### Induced Fit Model

An early analogy that illustrates the relationship between an enzyme and its **substrate** (i.e., starting material) is a lock and key. Only specific substrates can bind and be modified by specific enzymes. While this analogy explains the specificity of enzymes and their substrates, it falls short of explaining the true activity of enzymes. The **induced fit model** does a much better job of explaining both characteristics.

According to the induced fit model, when a substrate molecule binds to the enzyme, the enzyme changes shape, and since the substrate molecule is bound by the enzyme, its shape is altered, too. The substrate molecule is thereby moved into a better position to transform into the final product, reducing the energy and time required for the chemical reaction to occur. You could compare this to cutting snowflakes out of a sheet of paper. If someone had to cut every side of the snowflake by hand, it would take much longer, require more energy, and likely not look very symmetrical in its final form. This method would be like a chemical reaction occurring naturally. However, if the paper is first folded and then cut, fewer cuts are required, the snowflake is produced faster, and all sides will be the same. Enzymes fold and then process substrates in much the same fashion.

## Glycolysis

Enzymes are essential for processing organic molecules and releasing the energy stored within chemical bonds so cells can use it. In this section, the biochemical process of breaking down carbohydrates (sugars) will be explained.

While **aerobic respiration**, or using oxygen to produce energy, requires a complex series of enzymatic reactions and intermediate molecules, students should follow the number of carbon molecules (either into intermediates or released as $CO_2$) to understand how energy is recovered from the carbon bonds of glucose. Complete aerobic respiration results in 6 $CO_2$ molecules being formed from 1 glucose (6-carbon sugar) molecule.

Although many organic molecules can release energy to cells, the principal molecule used to release energy is glucose. The catabolic process of breaking down glucose is termed **glycolysis**. Glucose is a 6-carbon sugar that is imported into the cytoplasm where glycolysis begins. This is also the first stage of either aerobic (using oxygen) or anaerobic (lacking oxygen) respiration.

Fermentation is how cells produce energy when oxygen is in short supply. In yeast, a byproduct of this process is $CO_2$ and ethanol. In animal cells, the byproduct is lactic acid, and when this accumulates in your muscles they are very sore the next day.

During glycolysis, glucose is processed through 10 enzymatic steps from a 6-carbon molecule into 2 separate, 3-carbon molecules called **pyruvate**. These much smaller molecules can be transported into mitochondria where the remaining stages of aerobic respiration will occur. In addition to pyruvate, electrons will be captured by coenzymes and energy is released as a result of glycolysis.

**Nicotinamide adenine dinucleotide** ($NAD^+$), a coenzyme found in animal cells, captures and transports electrons from one chemical reaction to another. By collecting electrons released during specific stages of glycolysis and combining them with hydrogen protons, $NAD^+$ becomes NADH. In this form, electrons are moved to where they are required. Additionally, as chemical bonds are broken and rearranged, some of the bonding energy is released and is used to convert ADP (adenosine diphosphate) into ATP (adenosine triphosphate), which is the energy currency of the cell. Thus, ADP is the building block upon which a third phosphate is added (to produce ATP) and contains the energy that the cell can use for many biological activities in the bond between the second and third phosphate. Thus, the products of glycolysis are 2 pyruvate molecules, 2 NADH coenzymes, and 4 molecules of ATP.

In the early stages of glycolysis, 2 ATP molecules are used to prepare intermediate molecules for the next steps of glycolysis. While 4 ATP molecules are produced during glycolysis, the net production of ATP is only 2.

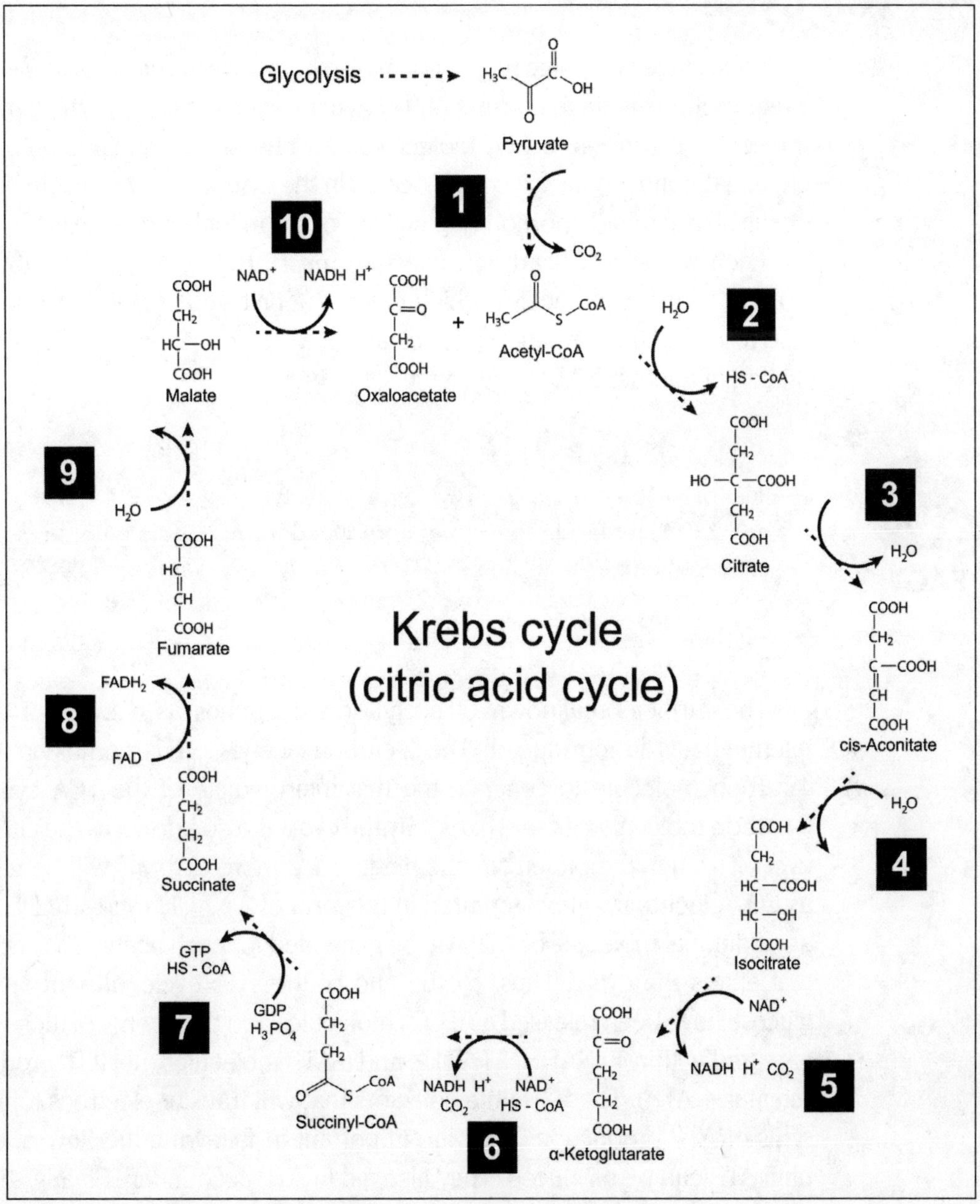

The metabolic process of breaking down glucose and capturing the energy contained within the chemical bonds between carbon atoms is illustrated in this biochemical chart.

## TCA Cycle

This next stage of aerobic respiration is also known as the **citric acid cycle** or the **Krebs cycle**. This stage consists of 9 enzymatic steps that break the 2-pyruvate molecules down into 6 $CO_2$ molecules and release the remaining energy in chemical bonds. While glycolysis occurs in the cytoplasm, the citric acid cycle occurs in the middle portion (**the matrix**) of the mitochondria.

Each pyruvate (3-carbon molecule) must first be converted into acetyl-CoA (a 2-carbon molecule), which enters the first step of the TCA cycle. This also results in the formation and release of a $CO_2$ molecule as well as the formation of another NADH coenzyme.

Just prior to entering the TCA cycle, 2 acetyl-CoA, 4 ATP, 4 NADH, and 2 $CO_2$ molecules have been produced from a single glucose molecule. Remember that the net ATP production thus far is only 2 ATP.

The further breakdown of acetyl-CoA continues as a cycle of carbon intermediates is introduced. The 2-carbon acetyl-CoA is combined with a 4-carbon molecule to generate the first intermediate of the TCA cycle, the 6-carbon molecule citrate. This is why the cycle is also known as the citric acid cycle. Through 8 more enzymatic reactions, 2 more carbons will be released as $CO_2$, electrons will be captured in the form of 3 NADH (and 1 $FADH_2$), and an additional molecule of ATP will be generated for each acetyl-CoA molecule that enters the cycle. Thus, by the end of the TCA cycle, all carbons from glucose have been released as 6 $CO_2$ molecules and the energy of those bonds captured within 10 NADH, 2 $FADH_2$, and 6 ATP molecules (net ATP production remains 4 ATP). These are the cofactors that will transfer electrons to the last steps of ATP production. They are important in following the flow of energy until ATP can be produced. They also aid in ATP accounting. Clearly, the bulk of the energy from glucose is contained within the coenzymes and has not yet been converted into ATP for the cell. That process of ATP formation also happens in the mitochondria as the electron transport system.

# Electron Transport System

To recover the energy from glucose that is contained within the coenzymes NADH and $FADH_2$, the electrons of these molecules are transported along a series of molecular complexes called the **electron transport chain**. This movement of electrons powers the creation of more hydrogen ions that will drive the generation of many ATP molecules.

This system is similar to a hydroelectric reservoir, where the dam is the inner membrane of the mitochondria and the water is the electrons. Just as water builds up in the lake and, due to gravity, has a large potential energy to flow through the turbines of the dam to create electricity, protein complexes in the mitochondrial membrane use the energy of the electrons from the coenzymes to move hydrogen ions into the intermembrane space. This reservoir of ions then flows back into the matrix of the mitochondria through the final protein complex, called ATP synthase. Just as the flow of water powers the turbine in the dam, the flow of hydrogen ions enables ATP synthase to generate ATP molecules.

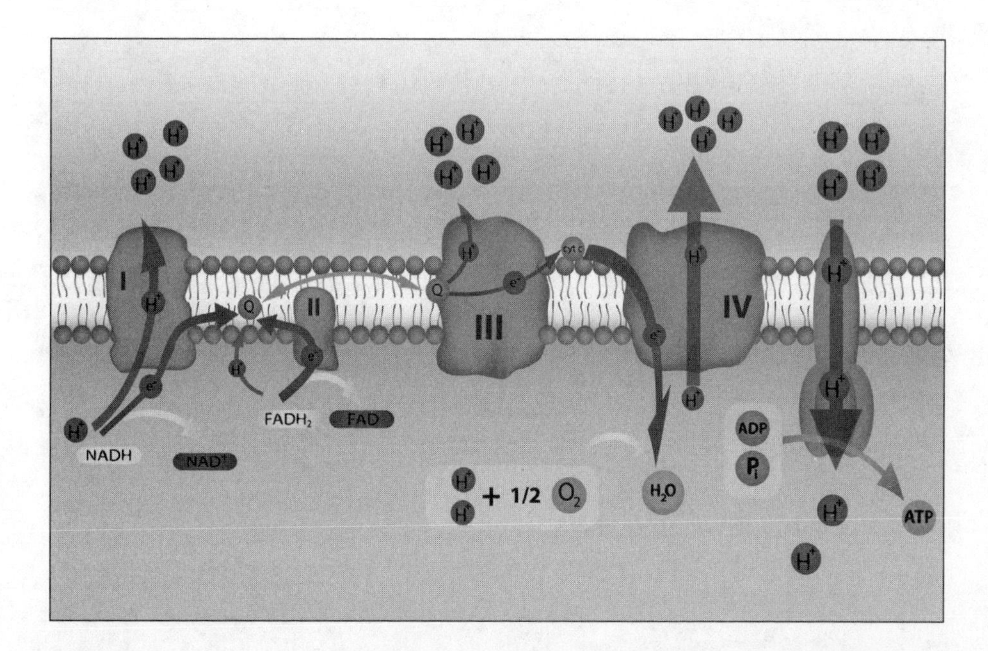


Electrons are passed from the cofactors generated in the Krebs cycle to the proteins of the electron transport chain. This enables the movement of protons ($H^+$) into the intermembrane space of the mitochondria where they accumulate until they flow by diffusion through the final protein in the process (ATP synthase), which drives the production of ATP (conversion of ADP into ATP).

Although only a net of 4 ATP molecules were produced during the stages leading up to the electron transport chain, the energy bound in the coenzymes as electrons from glucose is used to generate an additional 32 molecules of ATP. Finally, the tally of products at the end of aerobic respiration is 36 net molecules of ATP and 6 molecules of $CO_2$ for each molecule of glucose that enters this metabolic pathway. Without adequate fuel (glucose) or sufficient oxygen to finalize the process, energy cannot be produced in enough quantities to keep the body functioning in an active state. For humans, this importance is illustrated in a fight-or-flight situation, where your body automatically gears up for high activity. Without energy, this physical reaction would not be possible and the chances of your survival would be greatly reduced.

CHAPTER 4

# Human Tissues

The human body is a complex machine that is made up of many different but interrelated parts that must all function together to keep an individual healthy. Because the human body is so complex, biologists have categorized its parts into smaller units in order to learn more about the complexity of each. Just as the Earth consists of oceans and continents and each continent contains many countries, which themselves are composed of several states containing smaller cities and towns, so does the human body consist of several organ systems (digestive system, respiratory system, etc.). Different organs function together to form such systems. For example, the stomach and intestines are some of the organs that make up the digestive system, and further distinction breaks organs down into the tissues of which the organ is made.

# Tissue Origins and Development

Tissues are collections of similar cells that relate an essential function to the organ. While the adult human body consists of over 200 different cell types, each human began her life as a single fertilized egg cell, which divided and gave rise to all the rest of her cells.

Early in development, the newly forming cells produce 3 distinct layers of cells from which all the cells of the body are derived. These **germ layers** (foundational embryonic tissues) are called the **ectoderm**, **mesoderm**, and **endoderm**. The ectoderm (*ecto* meaning "outside") refers to the tissue that makes up your skin, which is the outer covering of the body. The human body is often considered a tube within a tube. In this sense, the ectoderm produces the outside tube of skin and the endoderm (*endo* meaning "internal") will produce the tube on the inside. The digestive tract makes up the inside tube of the human body. This leaves a lot of tissue and organs in between, which make up the mesoderm (*meso* meaning "middle"); muscle, bone, blood, and connective tissue are all derived from this middle "stuff" that is produced early in cell development.

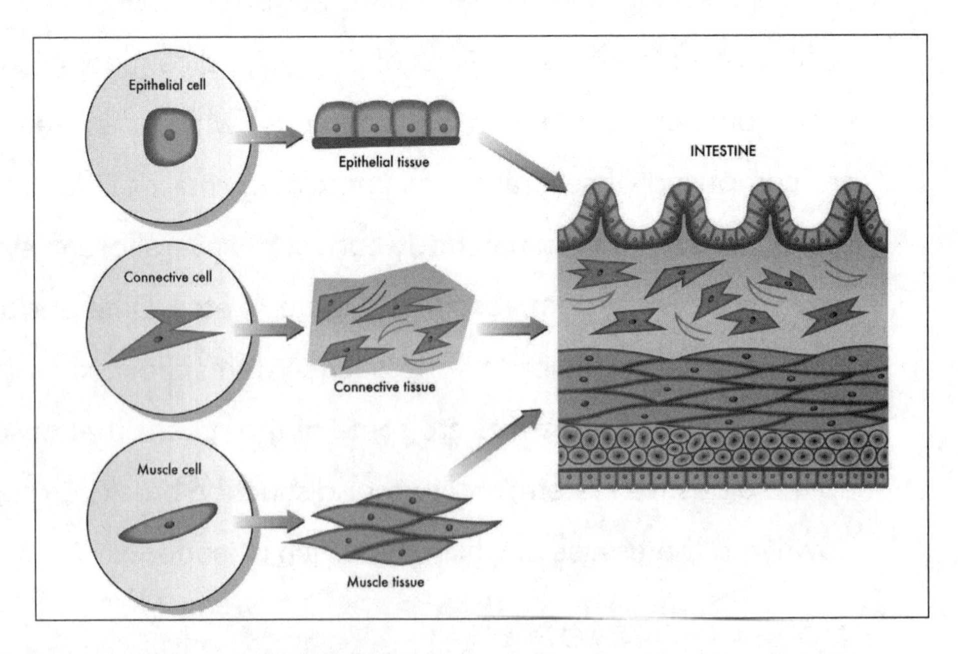


In this simplified illustration of the epidermis, the epithelium is shown as a single layer of cube-shaped cells. Thus, this type of epithelium is classified as simple cuboidal epithelium.

# Epithelial Tissue

**Epithelial tissue** covers a surface or forms the lining of a hollow organ. It is the first of the four types of tissue that make up the human body. For example, the surface of your skin is an epithelium, as is the inside lining of your stomach and intestines. In fact, the inside of your body cavity is covered with a thin layer of epithelial cells. The function of these covering cells is to protect and to provide a watertight barrier to keep material out (e.g., skin preventing pathogens from entering) or keeping material in (e.g., lining of the stomach keeping hydrochloric acid from damaging other areas of the body). When you consider all of the different areas in which epithelial tissue contributes to organs, it is not surprising that these cells must come in many different shapes.

**Epithelial cells** are classified, in part, based on their shape. Cells can be very flat, much like a fried egg, with the nucleus of the cells bulging upward like the yolk of the egg. These flat cells are classified as **squamous epithelium**. Epithelial cells can also be present as cube-shaped cells where the width of the side is the same as the height of the cell. These cells are known as **cuboidal epithelium**. Lastly, cells that are taller than they are wide, thus looking like columns, are classified as **columnar epithelium**.

## Number of Layers

Epithelial tissue composed of a single layer of cells is called a **simple epithelium**, whereas multiple layers of these cells are referred to as **stratified epithelium.** In this stratified layer, only the cells at the bottom of the layer are in contact with the underlying tissue. This is different in some areas of the body where the epithelium appears to have multiple layers, but all of the cells are in contact with the bottom of the layer. In these areas, the cells don't all reach the free surface, and this forces the nuclei to be present at different distances from the surface, creating the stratified or layered look. This type of epithelium is classified as **pseudostratified** and is easy to confuse with the truly stratified layers in which there are distinct layers of cells stacked one on top of the other.

When describing the epithelium, both its shape and the number of layers are considered. A single layer of flattened cells would be called a **simple squamous epithelium**. Likewise, an epithelium consisting of multiple layers and having a surface layer of cube-shaped cells would be called a **stratified cuboidal epithelium**.

When multiple layers are present, the shape of the cells at the surface of the epithelium tissue will be used in the classification regardless of the underlying cells.

**Transitional epithelium** is the last classification based on layering and shape. This tissue is found along the urinary tract in places that are in contact with urine, and in the lining of the bladder that must stretch when filled. While this is a stratified epithelium, the surface cells are large and either dome shaped (when the bladder is empty) or flattened (when the bladder is full). Often, the cells of this tissue will contain 2 nuclei that makes for easy identification.

## Apical Modification

The top of epithelial cells that line the lumen, or the inside of a hollow organ, is referred to as the **apical surface**; these cells have membrane specializations that affect the physiological function of the tissue. One such modification that may be found on the epithelial cells lining the respiratory tract is **cilia**. These "hairlike" structures extend upward from the **apical cell** surface and are capable of bending and moving back and forth to move materials within the organ. The cilia will move when a core of microtubules and microtubule-associated proteins pull and bend the microtubules in unison.

On cells in the intestines, the apical modifications are called **microvilli**. While these "fingerlike" projections extend upward into the lumen, they are not present to move materials. Rather, they are there to increase the surface area of the cell for greater absorption of nutrients and water. These extensions possess a core of **microfilaments** (such as actin), which produce a more rigid and less flexible increase in the apical surface.

## Basement Membrane

Beneath every epithelial layer is a zone of molecules that aids in anchoring the cells to the underlying tissues in much the same way as a foundation or a basement secures a home to the ground. The **basement membrane** (BM) is also a transition zone where cells anchor to molecules such as **laminin** (a cell-adhesion molecule of the basement membrane), and other BM molecules interconnect with the underlying connective tissue. This interlocking zone firmly anchors the epithelial layers to the fibrous connective tissue.

Together the lamina lucida and lamina densa compose the **basal lamina**. Some people confuse this with the basement membrane, but they are not interchangeable terms.

The BM consists of 3 zones each referred to as a **lamina**, which is another term for layer. The **lamina lucida** is a clear layer directly beneath and in contact with the bottommost epithelial cell. This is the layer containing the cell-adhesion molecules. The **lamina densa** is next. This layer gets its name because it's darker when compared to the lamina lucida. The lamina densa is darker because it is a highly compact network of **type IV collagen fibers** that resemble a net. This provides another anchorage point for the cells.

The deepest layer of the BM is the **lamina reticularis**. In this layer, fibers from the underlying connective tissue extend upward and interconnect with the molecules of the lamina densa. When considering the basic plan of the human body, many view it as a tube within a tube design. The outer tube is the skin and the inner tube is the digestive track. Thus, human beings are two tubes of epithelium with some pretty important tissue in between.

## Connective Tissue

As the name implies, connective tissue (CT) joins other tissues together, such as epithelial tissue tied to a muscle or epithelial tissue tied to nervous tissue. Connective tissue is composed of cells and molecules that function together for this adhesive process. They are categorized based on the density of the connective tissue as well as the proportion of cells to fibers within the tissue.

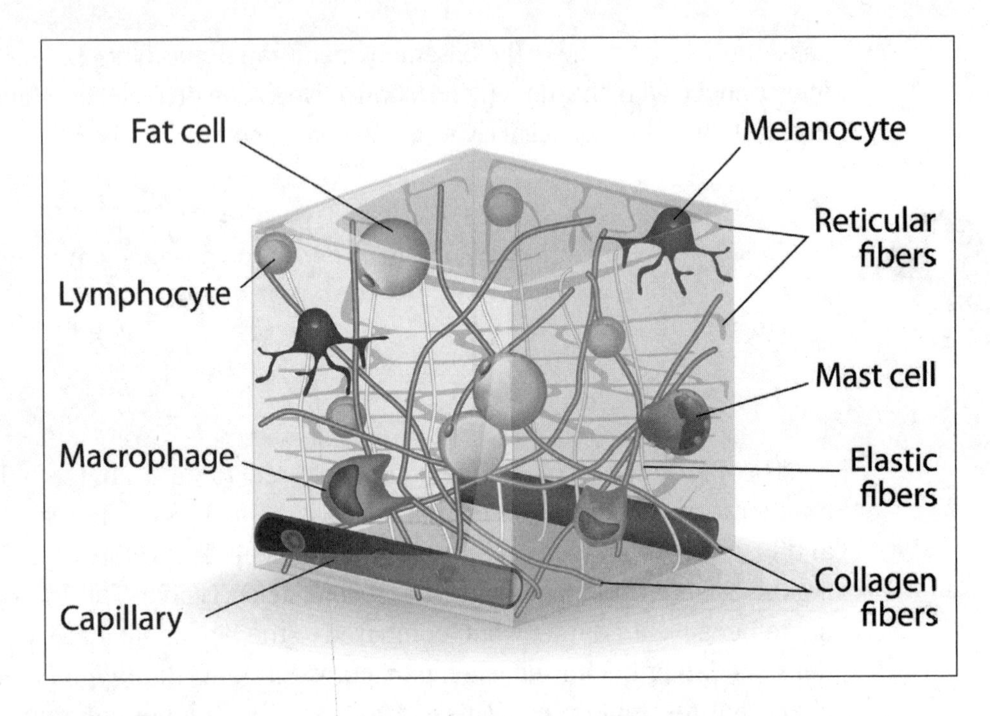


Connective tissue is composed of diverse collections of cells and fibers that link other tissues together into organs.

## Cells

**Fibroblast** is the principal cell of connective tissue. This cell deposits fibers that are found in all connective tissues: **collagen**, a protein resistant to stretching and therefore able to give the tissue tensile strength. These cells also produce elastic fibers that allow the tissues to rebound after being stretched, such as blood vessels that stretch when the heart beats and rebound when the heart relaxes. **Macrophages** are also found in connective tissues and function as the vacuum cleaners of the body by removing pathogens and debris from the area. Immune system cells such as **mast cells**, which produce inflammatory molecules, and **plasma cells**, which produce antibodies, are connective tissue cells.

Fat cells, or **adipocytes**, may also be present in connective tissue. These cells are mostly droplets of fats (lipids, cholesterol, or fatty acids) that are stored and released depending on the body's available energy. When fuel for the body is abundant, materials are stored as fat. In times when fuel is scarce in the blood, fat is converted into a transportable form that is shipped back into the blood to be used in the body to produce energy.

Fat cells will grow either larger or smaller depending on the amount of fat stored within them. Your body will not significantly increase or decrease the number of fat cells in your body. The only way to decrease the number of fat cells is through liposuction.

## Classification

Connective tissue is classified based on the percentage of cells to fibers and how tightly packed the fibers are within the connective tissue.

- **Loose connective tissue**, as the name implies, consists of widely spaced fibers and many cells migrating within the open spaces. This provides highly accessible tissue that immune system cells use to enter to fight an infection. Areas where you find loose connective tissue include the tissue around large blood vessels, beneath the epithelium of the skin, and around the digestive and respiratory tracts.
- **Dense irregular connective tissue** has many more fibers and fewer cells than loose connective tissue and is subclassified based on the orientation of the fibers. For example, in the dermis of the skin, the collagen fibers have a swirling and disorganized pattern that results in the classification of dense irregular connective tissue.
- **Dense regular connective tissue** is made up almost entirely of collagen or elastic fibers and contains few cells. Since the fibers are tightly packed and arranged parallel to one another, this tissue is classified as dense regular connective tissue. Ligaments and tendons that connect bone to bone or muscle to bone, respectively, are resistant to stretching and therefore are composed of collagen fibers with almost no spacing in between.

# Muscle Tissue

**Muscle** is not only the engine that moves the body, it also moves materials through the body. All muscles have one job: to contract. Muscle contraction can only happen with the sliding action of 2 proteins, **actin** and **myosin**.

Therefore, regardless of their appearance and classification, all muscles possess an abundance of these 2 contractile proteins.

## Actin and Myosin

For all muscles to contract, the overlapping actin and myosin molecules must slide toward each other, pulling each end of the cell and shortening the muscle. To envision the arrangement of these contractile proteins, extend your hands and touch the fingertips of your middle fingers together so you can view the palm of each hand, then slide just the fingertips of one hand between the fingertips of the other. This would resemble the arrangement of actin (fingers on the right hand) and myosin (fingers on the left hand) when a muscle is relaxed. During a contraction, these molecules will slide along each other much like your fingers sliding together until the tips of each finger can move no closer to your hand. This would now look like a set of fully contracted muscle actin and myosin molecules. Notice that when contracted, your wrists are closer to each other, much like the ends of a muscle would be closer when fully contracted (shortened).

## Skeletal Muscle

Skeletal muscle, as the name implies, is muscle that is attached to bones. This arrangement allows the muscles, which are the engines, to move the bones, which are the levers upon which actions can occur and work can be done. Pulling along a single plane, skeletal muscle cells are arranged into long cylindrical cables that are very well suited for strong contractions. During embryonic development, individual muscle cells fuse together to create long tubes of muscle cells, which then contain many nuclei. These are called skeletal muscle fibers. All of the nuclei are pressed to the periphery of the cell membrane because the middle of each muscle fiber is filled with long columns of overlapping actin and myosin molecules. The repeating and overlapping nature of the actin and myosin give skeletal muscle cells a rather striated appearance.

The repeating units of actin and myosin, which are arranged in series like the links in a chain, are called **sarcomeres.** Although each sarcomere shortens only a slight amount, all sarcomeres shortening at the same time results in the entire muscle organ shortening up to a few centimeters.

A functional difference between the 3 muscle types explained in this section is that skeletal muscle is the only type of muscle in the human body that is under voluntary control. Cardiac and smooth muscle are involuntary muscle types. Picking up a glass or running a race could not occur without your conscious control. Although some skeletal muscles could be under involuntary control, such as the diaphragm during hiccups, this can be overcome with conscious thought (e.g., holding your breath). But if you do so long enough and lose consciousness, your involuntary control would resume.

## Cardiac Muscle

Like skeletal muscle cells, cardiac muscle also contains overlapping actin and myosin arranged into sarcomeres, yielding a striated appearance. However, there are some clear differences between skeletal and cardiac muscle. All cardiac muscle is under subconscious control and is considered involuntary. While you can increase or decrease your heart rate by your level of activity (e.g., running can increase the rate, while lying down will decrease the rate), heart rate cannot be voluntarily controlled by thought itself.

Another difference between cardiac and skeletal muscle cells is that cardiac cells are branched at many different points, unlike the linear tubes of skeletal muscle. These branch points will link cardiac muscle cells together into laminae (interwoven layers or sheets). Laminae allow the 3-dimensional contraction of the heart rather than a simple linear contraction as in skeletal muscle. The attachment points between muscle cells are one of the most distinctive features of cardiac muscle and are called **intercalated disks**.

Intercalated disks allow the muscle cells to hold on to each other tightly while contracting so they don't pull apart. Additionally, these disks also contain membrane tunnels called **gaps junctions**, which allow the cytoplasm of one cell to flow unimpeded into the adjacent cell. This structural and functional interconnection results in cardiac muscle cells joined via gaps junctions contracting at the same time. This creates a **syncytium** (a single functional unit comprised of many parts working in unison) that permits the heart to function as a single unit although composed of many parts.

### Smooth Muscle

Unlike the other 2 types of muscles, smooth muscle lacks a striated pattern. It is named after its smooth appearance. This does not mean smooth muscle lacks actin and myosin. Instead, smooth muscle lacks the **sarcomeric** arrangement of the contractile proteins. In smooth muscle, the overlapping actin and myosin are attached to points on the plasma membrane called **dense bodies**, which are scattered all over the surface of the cell. This 3-dimensional pattern will cause the cell to collapse upon itself when contracted. Because of the various arrangement of the contractile proteins, the cell appears as one density with no bands.

Like cardiac muscle, smooth muscle is involuntary and may be joined together with gaps junctions to form belts or bands of smooth muscle tissue. These are found surrounding hollow organs, such as the digestive tract, urinary tract, and blood vessels, and assist in the movement of materials such as food, urine, and blood through the body. The belts of smooth muscle function as regulators of movement. For example, the pyloric sphincter is a belt of muscle between the stomach and small intestine that regulates when and if material passes from the stomach to the intestines. Contraction of this smooth muscle belt causes a narrowing and even closing of the passageway, and a relaxation leads to an open passage.

## Nervous Tissue

Nervous tissue is capable of sending electrical signals from one place to another in the body. This can either bring information to the central nervous system from the body for processing or send information out to the body from the central nervous system. The nervous system comprises many different parts. This section will discuss the cells that are involved in the formation of both the central (i.e., brain and spinal cord) and peripheral nervous systems.

## Neurons

Neurons are the signaling cells of the nervous system and come in a myriad of shapes and sizes. Generally, neurons have processes extending from the cell body called neurites, which make the cell look somewhat spiky. These are defined based on the direction the signals travel. For instance, if the electrical signal travels toward the cell body, the neurite is called a **dendrite**. If the signal moves away from the cell body, it is an **axon**.

Typically neurons will have many smaller-diameter dendrites and one longer, thicker axon.

## Neuroglia

While neurons are the signaling cells of the nervous system, they only make up 20 percent of the nervous system. The bulk is made up of the supporting cells of the nervous system that are collectively referred to as **neuroglia** or nerve glue. In much the same way as the movie stars on the screen are only a small fraction of the people involved in making a movie, the behind-the-scenes supporting members for the nervous system are the neuroglial cells. The neuroglia protects and provides nutrients to the cells of the nervous system.

### Myelinating Cells

One group of neuroglia has the capacity to insulate the axons of neurons. These cells will wrap their plasma membranes around the axon 40 or 50 times, much like an electrician wrapping electrical tape around a bare wire. The term **myelin** refers to the area of the membrane that is tightly wrapped around the axon (to **myelinate** something is to wrap around it). Myelin insulates the axon, and in doing so creates a much more rapid conduction velocity of the electrical signal. The neurons in the peripheral nervous system supplying areas of the body such as in your arms and legs will be myelinated by **Schwann cells**, which are responsible for wrapping peripheral nerves. Each Schwann cell will wrap a portion of its membrane around an axon and myelinate in one region of one axon. In the central nervous system, however, **oligodendrocytes** will myelinate axons. Much like an octopus with many arms, these cells extend several cellular processes to axons and can thus myelinate several different axons.

### Astrocytes

Another type of neuroglial cell is the star-shaped **astrocyte**. These cells protect the nervous system from infection by covering blood vessels with their cellular processes and regulate the movement of materials out of the blood stream and into the nervous system. As part of the blood-brain barrier, they screen for pathogens and assist in the transportation of nutrients as well as the processing of waste.

### Microglia

**Microglia cells** are the vacuum cleaners of the central nervous system. Known as **phagocytic cells**, they will patrol the nervous system and look for pathogens and debris and remove them through a process called **phagocytosis**, or engulfing the foreign object into their cell bodies. This way, microglia effectively eliminate these detrimental materials from the central nervous system before they can do any damage.

### Ependymal Cells

The last type of neuroglial cell present in the central nervous system is the ependymal cell. These cells produce **cerebrospinal fluid**, which flows around the spinal cord and the brain and functions as a shock absorber against blows to the head and spine. Additionally, as cerebrospinal fluid is produced and as it is eliminated, it creates a current that will also effectively assist in the blood-brain barrier because pathogens would have to swim across this river of flowing cerebrospinal fluid before gaining access to the central nervous system.

CHAPTER 5

# Skin

Skin is the largest organ in the human body. It's the external covering of the body and it serves to protect against infection and prevents the body from drying out. Composed of several layers, the skin can be classified as either thin skin or thick skin. Also present in skin is hair, which can detect touch and helps keep the body warm. To cool the body on hot days, sweat glands found in the skin will produce sweat that cools the body as it evaporates from the surface.

## Epidermis

While skin covers the entire body surface, all skin isn't created equal. The distinction between the skin on your arm and the skin on the soles of your feet is significant and largely based on the thickness of the **epidermal** (top) part of the skin.

Skin may contain up to 5 individual layers (**strata**) that make up the top layer, or the epidermis: stratum basale, stratum spinosum, stratum granulosum, stratum lucidum, and stratum corneum. Starting at the bottom adjacent to the basement membrane is the **stratum basale**, where cells divide and produce a continuous supply of new cells as the old cells are shed from the surface. The next layer as you move up is the **stratum spinosum**, which gets its name because its cells have a spiny shape when prepared for histology. **Melanocytes** are the pigmented cells in the lower layers that give skin cells their individual color.

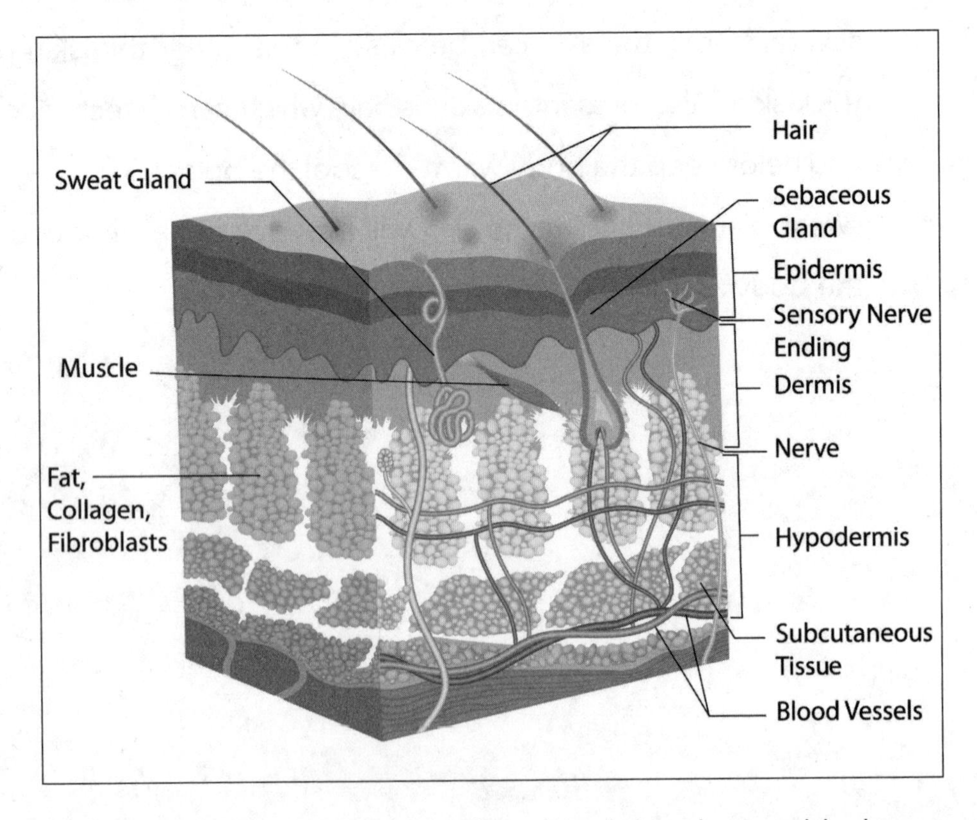


Skin is composed of the surface epidermis layer, the underlying dermis, and the deepest layer: the hypodermis.

The melanocytes occur throughout the spinosal layer, using their membranes and pigment granules to cover over and protect the dividing cells at the bottom from damage from ultraviolet radiation, kind of like an umbrella keeping you in the shade on a beach. As skin cells (**keratinocytes**) are moved higher and higher, they become cells of the **stratum granulosum**. These cells are filled with granules of **keratohyalin**, which give the cells a grainy appearance. This is also the last layer of living cells in the epidermis. The **stratum lucidum** is a thin translucent layer of the skin and is composed of dead skin cells. On top of this layer is the final layer, which is the most variable in its thickness. The **stratum corneum** is composed of multiple dead cell layers and is the first line of defense against infection for the body.

Cells are continually shed (in a process called **desquamation**) from the epidermal surface, because new cells are continually made. It takes a new skin cell about 14 days to move from the bottom layer to the surface of the skin.

Thick skin is found on the palms of the hands and soles of the feel. Of the 5 previously mentioned layers, the stratum corneum is by far the thickest layer. It provides great protection against the friction that occurs when walking or grasping objects; however, unlike the rest of the skin that covers your body, you will not find hair follicles or sebaceous glands in thick skin. Furthermore, you will find fewer sweat glands in thick skin than are present in thin skin. While your feet and palms may get moist, the volume of sweat produced is much lower than what is produced in other areas of the body.

The majority of the body is covered with thin skin (with thick skin being found on the palms of the hands and soles of the feet). The name "thin skin" stems from the thin stratum corneum layer (when compared to thick skin). In fact, the epidermis as a whole is thinner in this type of skin, since it lacks 2 of the 5 layers that are found in thick skin (i.e., stratum granulosum and stratum lucidum are missing). Another contrast between thin and thick skin is that thin skin possesses many hair follicles and sebaceous glands that support the growing hair. Additionally, while both skin types contain sweat glands, thin skin has a much higher density of these glands to aid in cooling the body.

# Dermis

The **dermis** (dermal layer) is the layer beneath the epidermis that forms a transition zone between the underlying connective tissue and the epidermal layer above. It consists of dense irregular connective tissue made up largely of collagen fibers, elastic fibers, and fat tissue. Within the dermis are blood vessels that support the skin, as well as many nerve endings and receptors that detect pressure, pain, and temperature.

Meissner's and Pacinian corpuscles as well as Ruffini's end organ detect pressure (touch) and are located in the dermis of the skin. These and other sensory receptors in the skin enable the detection of a wide variety of stimuli even if they are very mild such as the lightest touch or the slightest change in temperature.

# Hair and Nails

Human skin may also be modified into the structures known as hair and nails. Hair and nails are composed of the same material that makes up the surface stratum corneum of the skin. The major difference is how compact these layers are and how they are arranged with the other dead cells in these layers.

## Hair

Imagine taking the layer of dead cells that is the stratum corneum (surface layer of skin) and rolling it into a tightly wound tube of dead cells: That is a hair. The hair follicle is simply a deep pit on the epidermis that projects downward into the deeper dermis of the skin. Using this analogy, the walls of the follicle are synonymous with the stratum corneum. As this epidermal pit is pushed downward, the walls of the pit are the former stratum corneum of the skin, and thus form a tube of dead cells with the other layers of the epidermis surrounding and supporting the shaft of the hair as it grows. At the deepest part of the follicle is the hair bulb. This is where the cells divide. This is also where they are nourished by blood vessels that enter into the follicle and support the growth of the hair. As new cells are added to the hair, the hair is continually pushed out of the follicle and onto the surface of the skin.

## Nails

The surface of the skin is relatively soft. However, the nails, which are made of the same material as skin and hair, are drastically harder and consist of tightly packed layers of dead cells. What is commonly known as hair is in fact nail plates, which, like hair, are synonymous with the surface stratum corneum of the skin. Although they are made of the same material, nails are much harder than hair because the cells are packed much more tightly, which results in a more dense structure (nails). This hardened plate of dead cells remains attached to the underlying epidermis via the nail bed, which tightly holds on to the plate to prevent the nail from falling off. At the base of the nail, you'll find the cuticle (**eponychium**). The cuticle is a portion of the skin's epidermis that overlaps the newly forming nail plate as it moves forward. Beneath the cuticle new nail plate material is continuously being produced and pushing the nail forward.

At the end of the nail, the plate extends beyond the tip of the finger, forming a crevasse, which is great at collecting dirt and is technically called the **hyponychium**. This is the part of the nail that you are most familiar with and must trim on a regular basis to prevent the nails from growing too long.

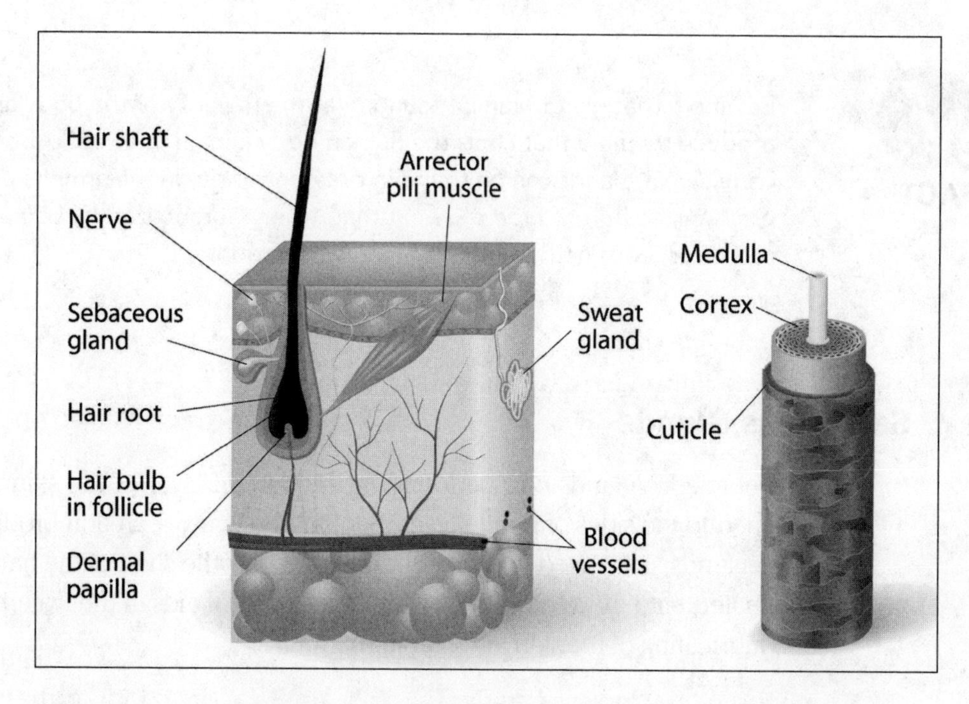


Hair follicles extend deep into the dermal layers of skin and are composed of the same layers that make up the epidermis. The cuticle, cortex, and medulla of the hair are compacted dead skin cells.

## Sweat and Sebaceous Glands

While the skin is an intact (contiguous) sheet of cells that protects the human body, there are accessory structures that enable the skin to cool the body as well as condition the skin itself. These important glands are the sweat glands and the sebaceous glands.

### Sweat Glands

In ancient times, Egyptians would hang wet linen cloth in doorways and windows to cool the inside of their living areas. As the water evaporated, it cut down on the heat in the air and effectively cooled the rooms. The human body uses this same approach to cool down. This is called sweating. When the surface of the skin is wetted, evaporation will occur, and as it does, your body gives off heat and your body temperature is lowered. The secretory portion of the sweat gland resides in the deeper parts of the dermis and produces the salty, protein- and lipid-rich fluid we call sweat. This is passed through long, coiled ducts upward and through the epidermis to be released onto the surface of the skin.

FACT

**Eccrine** sweat glands can be found over the majority of the body and produce the fluid that cools the human body known as sweat. **Apocrine** sweat glands can be found in greater density in the armpits and produce a different type of sweat that, when metabolized by bacteria, produces a distinctive and often unpleasant odor.

### Sebaceous Glands

Sebaceous glands can be found in the dermal layer of the skin and are attached to the sides of hair follicles. Sebum, the waxy secretion produced by these glands, is injected into the hair follicle, coats the shaft of the hair, and is eliminated onto the surface of the skin. This material aids in the waterproofing and lubricating of the skin and hair in mammals.

During the last stages of fetal development, the sebaceous glands produce a thick waxy layer called vernix caseosa that covers the fetus and protects the skin from the amniotic fluid in the womb.

## Wound Healing

An intact layer of skin is critical in preventing infectious agents from gaining access to deeper regions of the body, including the blood stream. Therefore, when an injury to the skin occurs, rapid and complete repair or closure of the wounded area must be accomplished. Because the skin consists of several layers, all of them will be involved in the repair process.

### Inflammatory Phase

The body's initial response to a wound is to take steps to minimize blood loss. Damaged blood vessels will reflexively constrict to slow the flow of blood to the damaged area. At the same time, **platelets**, or cell fragments in the blood that are responsible for initiating a blood clot, will activate and begin the process of clotting. They do this by forming a platelet plug to prevent the further loss of blood cells and to decrease the loss of blood plasma at the injury site. This plug will form the initial foundation of the full blood clot.

At the same time as the clotting, the body will also flood the injured site with body fluid and immune cells (white blood cells, or **leukocytes**), flushing the area of pathogens and destroying any pathogens that may remain. This **inflammatory** process is triggered nonspecifically as part of the immune system to protect and defend the body against a wide variety of pathogens.

### Proliferative Phase

As the wound site becomes filled with the blood clot and infiltrating white blood cells, connective tissue cells, or **fibroblasts**, will migrate into the area and begin to deposit a temporary scaffolding of connective tissue molecules to fill in the open gap in the skin. This material is the start of **granulation tissue** and will help close the wound and provide a foundation for the normal tissue constituents to regrow and re-form the original state of the skin.

Since the surface of your skin is an **epithelium** and not connective tissue, those epithelial cells on the edges of the wound will start to multiply and spread. They will then overgrow the granulation tissue that is forming in the wound. If the wound is too wide or the granulation tissue grows too extensively, then re-epithelialization will not occur and will result in a scar. Additionally, to provide nutrients to the newly growing tissue, blood vessels will grow into the granulation tissue and form new circulation for the freshly formed tissue.

Tissue reconstruction will continue as the wound is repaired and as the body attempts to close the wound. The blood clot will then start contracting and will pull the edges of the wound closer together, resulting in a smaller area to form new tissue.

Scars occur when there is an overproduction of connective tissue proteins such as collagen, which prevents the re-formation of normal epithelial tissue over the wound site. This often occurs if the wound site is too expansive for the normal epithelial tissue to repair.

### Maturation Phase

In the final weeks of wound repair, the last portions of granulation tissue will be removed by immune system cells of the body, which are called macrophages. All clotting material will be eliminated from the site via enzymatic and cellular action, and the new skin will be properly formed in the exact proportions and in the correct area. Following this period, most repairs of small wounds will result in skin that shows little to no sign of any injury or defect at all.

## Temperature Regulation

Most people would have little difficulty stating that the skin protects our bodies from disease and from dehydration. However, most are only peripherally aware of the role the skin plays in temperature regulation of the body.

## Sweat

As discussed earlier, when the human body is hot, it will sweat. As the liquid on the surface of your skin evaporates, heat is liberated from the skin and dispersed into the air, thereby cooling the skin. In this way, the skin and the associated blood vessels that supply the skin can be thought of as a radiator or a heat exchanger between the human body and the environment.

## Conserving Heat

The skin and its vascular supply also play a critical role in conserving body heat when the core body temperature begins to decrease on very cold days. When exposed to the cold, the skin of the hands and face will initially flush and turn red, which indicates the dilation of the blood vessels near the skin surface in an attempt to bring more of the body's core temperature to the skin and keep the tissue warm.

This attempt to warm the body will only continue to a point. When the body's core temperature falls enough, the blood supply to the skin will be restricted as the blood is redirected to deeper areas of the body and away from the skin and extremities. In this way, the blood doesn't approach those superficial areas, such as wrists and hands, where high densities of blood vessels are subject to losing heat quickly. This is the body's way of sacrificing less important areas to keep the core of the body's temperature from falling too low.

Frostbite is where the blood supply is completely shut off from the extremities and the tissue dies (**necrosis**), hence the blackened color that frostbitten tissue takes. Depending on how cold it is, exposed skin can become frostbitten in as quickly as 5 minutes.

# Diseases and Disorders

At the time of birth, a baby's skin is soft, smooth, and often completely clear of defects. Newborns may exhibit birthmarks or strawberry marks, which are often seen as damaged areas caused during pregnancy. However, as humans age and their skin is exposed to longer periods of ultraviolet (UV) radiation and environmental factors, the skin will likely exhibit localized changes in coloration or in the proliferation of cells. This can result in moles, freckles, or tumors.

## Acne

**Acne vulgaris** is a common skin disease that often begins or intensifies during adolescence and puberty because of hormonal changes that result in a greater production of oils from glands of the skin. Acne vulgaris arises when hair follicles and sebaceous glands become clogged and infected. These often manifest as pimples or pustules and will continue to occur until the underlying causes of the skin disorder (typically an increase in oil or sebum production) are relieved.

Another common problem associated with acne is the appearance of blackheads, or **comedones**. While these occur as a result of dirt or material blocking the hair follicles, they may become whiteheads or pimples if they become infected.

## Blisters

**Blisters** are common disorders that occur on the hands or the feet because of friction from walking or running in poorly fitting shoes or from using hand tools for long periods of time without gloves. These frictional forces tear and shear the cells of the **stratum spinosum**, causing a separation of cells; the plasma from blood vessels fills these voids and raises the upper layers of the epidermis. If the friction is continued and more damage occurs to the tissue, the blister may fill with blood due to vascular damage within the layers of the skin.

## Nevi

Permanent and benign colored areas of the skin, which are often called birthmarks, are termed **nevi** (nevus is singular). When this tissue is caused by a proliferation of **melanocytes** (pigment cells), then the tissue will have a brown to black coloration. If the colored tissue is more of a red color and is caused by a collection of blood vessels close to the skin, then this is a **vascular nevus** (hemangioma), or more commonly called a strawberry mark.

Moles are a form of melanocytic nevi. Freckles (**ephelides**) are flattened accumulations of melanocytes in a specific area typically brought about by exposure to UV radiation.

## Skin Cancer

**Basal cell carcinoma**, the most common form of skin cancer, is caused by out-of-control growth of the cells at the basal layer of the epidermis. If control of basal cell division is lost, due to genetic mutations that may be caused by dangerous environmental agents (such as cigarette smoke) or ionizing radiation (UV light), cancer occurs. Fortunately, this is a fairly slow-growing cancer and rarely spreads to other parts of the body. That being said, it can lead to disfiguring lesions if not treated.

**Melanoma**, while not the most common form of skin cancer, is the most dangerous. The pigment cells of the skin (melanocytes) lose control of their division and form colored tumors on the skin that may resemble moles (or form from moles). If you have any new pigmented and rapidly growing or changing area of your skin, you should see a physician as soon as possible. What makes melanoma so dangerous is that it may quickly spread widely throughout the body where it becomes difficult to treat.

**Squamous cell carcinoma** occurs when the most superficial layers of the skin divide out of control. These tumors may appear as red patches of flaky skin or open sores, and may bleed from or near the tumor itself. Like melanoma, these tumors are mostly caused from UV exposure and damage to the cells and can be disfiguring and dangerous if not removed before the cancer can spread.

CHAPTER 6

# Skeletal System

The bones of the body are the structural architecture that gives the human body its distinctive shape. Without the skeletal system, you wouldn't be able to do many of the things you can today. Walking, running, throwing, grasping, and chewing would all be impossible without bones. However, bones do much more than maintain shape and enable your body to do work.

## Bone Functions

You might be most familiar with bones found in Halloween decorations or on display in a museum where they mark the historical record of dinosaurs. However, it's this unseen organ that facilitates much of what the human body is capable of performing. Bones not only partner with muscles to move body parts, they also help with new blood cell formation as well as act as a storage compartment for calcium.

### Levers for Movement

If you have ever used a crowbar, you have demonstrated the concept of how levers work. Many bones in the skeletal system act as levers for the body, while the skeletal muscles (named such for their attachment to bones) provide the power for the work to be completed. Therefore, just like a lever system, the points where the skeletal muscles attach to the bones, and the particular bones that the muscles are attached to, dictate the power and strength required to accomplish a task.

Skeletal muscles will contract and pull in a straight line. One attachment point will be relatively fixed whereas the opposite attachment point will be moveable. The biceps muscle, for instance, has a relatively fixed attachment point (or **origin** of the muscle) at the scapula on the shoulder. At the opposite end of the biceps, the moveable attachment point (or **insertion** of the muscle) is to the radius bone of the forearm. When the muscle contracts, it pulls on both the origin and insertion, attempting to shorten the muscle toward the middle. However, because the shoulder is fixed, the only appreciable movement that can occur is for the forearm to be pulled closer toward the shoulder (i.e., flexing of the forearm). This "action" of the muscle, which is driven by the muscle contraction, is in fact a result of the involvement and arrangement of the bones of the skeletal system.

### Hematopoiesis

After just a few weeks following conception, the developing **embryo** will have grown too large for oxygen to diffuse to all the cells of the body. This is when red blood cells are formed from embryonic precursors and a primitive circulatory system is established. As the embryo grows and becomes a fetus, the spleen and the liver will serve as the location for new red blood cell formation.

However, these organs serve other functions for the adult human. Their blood cell production duties are focused on the long bones for the remainder of the developmental period and for the rest of the life of humans.

**Hematopoiesis** is the production of blood cells from embryonic stem cells present in the **marrow**, in the inner portion, of the long bones. **Erythropoiesis** is the specific production of red blood cells, while **leukopoiesis** is the production of white blood cells.

## Calcium Storage

Bone is principally composed of a hard, inorganic calcium phosphate matrix. While this gives bones the strength to resist gravity and support the movements of the body, it also gives the body a great reservoir of calcium. Unlike **enamel**, the calcium phosphate matrix of the teeth, bone is porous and filled with living cells that can repair and remodel bone on a continual basis. Therefore, as metabolic or dietary deficiencies in calcium occur, in order to provide calcium to the body (which is required for muscle contraction and nerve signaling) the body will rob calcium from the bones.

Osteoporosis is a disease of the elderly, particularly females, in which calcium and bone matrix is progressively lost. This weakens the bone and creates holes or pores (hence porosis). This weakening of the bones makes them much more susceptible to breaking.

Just as sugar levels in the blood are regulated by 2 hormones (i.e., insulin decreases blood sugar and glucagon increases blood sugar), calcium is also closely monitored and regulated by 2 hormones. **Parathyroid hormone** (parathormone) functions to increase blood calcium levels while **calcitonin** (produced by parafollicular cells of the thyroid gland) decreases blood calcium levels.

# Axial Skeleton

There are more than 206 bones in the adult body (some of which are formed by the fusion of several smaller bones in a newborn, which has approximately

270 bones), all of which are grouped into 2 main categories. First, there is the portion of the skeletal system that is aligned with the **vertical plane** or axis of the body. This is part of the axial skeleton and includes the skull, vertebral column, and rib cage.

## Skull

The skull, or **cranium**, is composed of numerous small and flattened bones that encase the brain, provide a base upon which the brain sits, provide attachment points for the upper and lower jaw, and form attachment points for connection to the vertebral column. Following are the different parts of the skull.

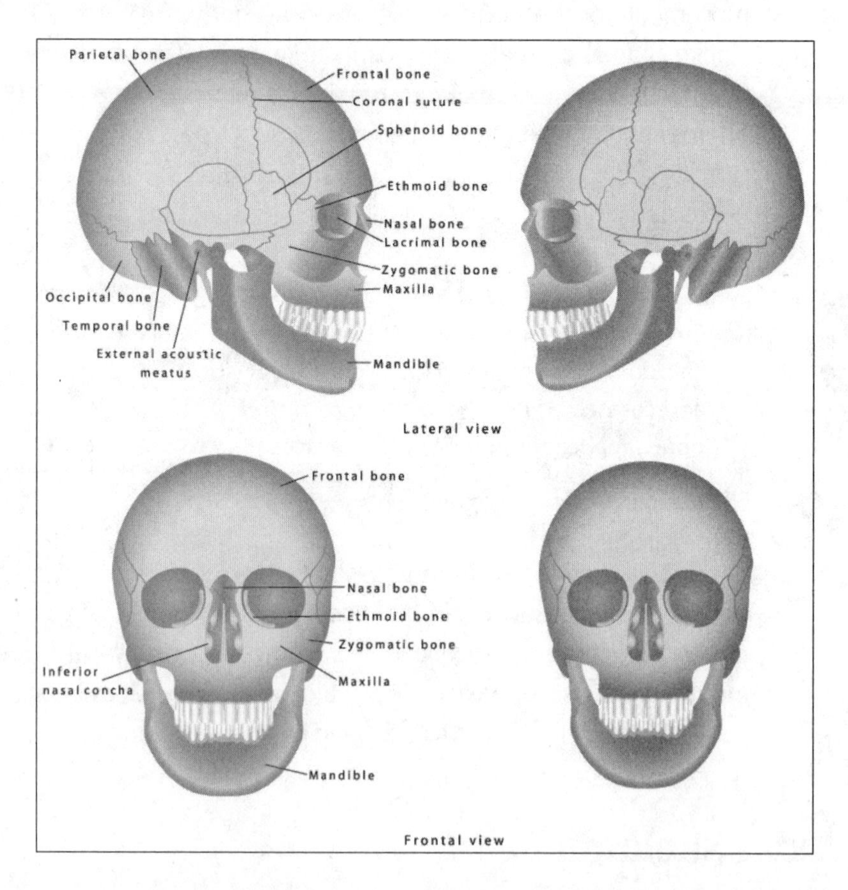


The human skull is constructed of numerous bones that encase and protect the brain, and form the upper and lower jaws.

### Neurocranium

The skullcap (**membranous neurocranium**) forms the roof of the skull. Many refer to this portion also as the **calvarium**. This is made up of several bones, which, at birth, are loosely joined together with spaces called **fontanelles** between them that allow for the rapid growth of the brain and later fuse into a single composite structure.

The "soft spot" on the top of a newborn baby's head is actually the largest of several fontanelles (points that the skull bones will grow into later).

The skullcap is made up of left and right frontal bones (the forehead), right and left parietal bones (the top and most of the side), and an occipital bone (back lower portion of the skull). Around the area of the ears are additional bones on the right and left that make up the temporal bones. These will allow passage of the auditory canal, 7 cranial nerves, and major blood vessels to enter and leave the skull.

### Endocranium

This is the base or the floor on which the brain sits. Imagine the skullcap as the roof and the walls for the brain and the **endocranium** as the floor. This is also referred to as the **chondrocranium** in lower vertebrates, such as the shark. In human development, these bones will first form as cartilage, and will later be replaced by bone in the process called **endochondral osteogenesis** (covered in later sections). As with the calvarium, the base is a composite structure that will grow and fuse into a single functional base for the brain. In some anatomy texts, the lower portion of the **occipital** bone, especially the part that surrounds the opening through which the brain stem passes (foramen magnum), is referred to as part of the endocranium.

### Facial Skeleton

The final portions of the skull are the bones that make up the jaw and the face. Often referred to in anatomy class as the **viscerocranium**, this section consists of the upper jaw (**maxilla**), lower jaw (**mandible**), and the bones of the nose and the palate of the mouth (nasal and palatine bones, respectively). Additional small bones are also present in the eye sockets and the deeper divisions of the nasal cavities.

## Vertebral Column

This supporting column for the human body, and also the protector of the spinal cord, is a highly diverse and composite structure consisting of 33 **vertebrae**, or the bones of the backbone, that are specialized depending upon their position in the body (i.e., neck, chest, hips, etc.). Each possesses a main body, the **centrum**, and is used in an interlocking fashion to hold each successive vertebra in place relative to its neighbor. Also, connection points are present on the vertebrae in the chest region where the ribs are attached.

The most superior (closest to the head) 7 vertebrae are called the **cervical vertebrae** and are often designated as C1–C7. Connecting the vertebral column to the **occipital condyle**, the boney extensions on the base of the brain, is the function of C1 (atlas). The next vertebra, or C2, is called the axis, and it allows the pivoting of the axis on the vertebral column. These are the only 2 named vertebrae in the body.

The next 12 vertebrae are the **thoracic vertebrae** (T1–T12). Each of these vertebrae projects away from the vertebral column at about a 45° angle and will form the connection points with the base of each rib.

Lastly, **lumbar** vertebrae (lower back) consist of the next 5 vertebrae (L1–L5) and are found in the lower portion of the abdominal area.

The remaining 9 vertebrae are actually fused into 2 units, 1 of which functions as a portion of the pelvic girdle (hip). The sacrum forms from the next 5 vertebrae (S1–S5), while the last 4 small vertebrae are fused together into the structure known as the **coccyx**, or tailbone.

## Ribcage

Consisting of the ribs, interconnecting cartilages, and the bones of the **sternum** (breastbone), the ribcage is the protective structure surrounding the vital organs of the thoracic region including the heart and lungs. In addition, the ribcage also helps you breathe.

### Ribs

These 12 pairs of curved bones reach from the vertebral column and connect to the ventral (chest) surface of the **thorax** via the sternum. The ribcage as a whole is capable of moving upward (and expanding) each time a person inhales. Likewise, when the breathing relaxes, gravity pulls the ribcage downward (and inward) to its returning position and a smaller volume, thus forcing the air in the lungs to be exhaled.

The first 7 most superior ribs (closest to the head) are referred to as the "true" ribs. While the composition of all ribs is identical, these first 7 are attached to the sternum via individual cartilage extensions called **costal cartilages**. The remaining 5 pairs are called false ribs; the first 3 pairs of these are indirectly attached to the sternum via cartilages that join the costal cartilage of the seventh rib, and therefore do not have their own costal cartilage. The final 2 most inferior (lowest) false ribs are called "floating ribs" because they are only attached to the vertebral column on the dorsal (upper) side of the body, leaving the ventral ends of the ribs unattached.

### Sternum

The sternum is a composite flattened plate of bones that defines the ventral surface of the thoracic region of the human body, or, more simply put, your chest. The middle portion of the sternum, and the bulk of its mass, is the body of the sternum. For the superior portion, the **manubrium** is a broader square-shaped bone that is connected to the body of the sternum and extends upward toward the neck. Finally, the superior portion of the sternum is formed by an arrow-shaped bone called the **xiphoid process**.

# Appendicular Skeleton

While the bones of the axial skeleton define the vertical axis of the body, the appendicular skeleton makes up the arms and legs, as well as the girdles that secure the limbs to the axial skeleton.

## Pectoral Girdle

The bones that support the arms are the **scapula** and the **clavicle** and compose the **pectoral girdle**. Also known as the collarbones, the clavicles are slender bones that attach and anchor the scapula to the manubrium of the sternum. The clavicle also acts as a platform for the attachment of muscles from the arms, chest, and the back.

The scapula (shoulder blade) is the broad and flattened bone that is clearly visible on the back. It projects from the shoulder toward the spine. The broad flattened surfaces of the scapula are locations for the attachments of large muscles of the back including the **supra-** and **infraspinous** muscle, as well as the **subscapular** muscle. The scapula is somewhat triangular in shape, with the base toward the spine and the point toward the shoulder. The surface of the point is actually a concave depression that serves as a location for the "ball" of the upper arm bone (**humerus**) to form a freely moveable joint with the scapula. This connection is stabilized by the ligaments and tendons of the shoulder joint.

Ligaments are connective tissues that attach bone to bone and resist stretching. Tendons are made from the same material as ligaments, but connect muscle to bone.

## Arms and Hands

Human arms are made up of only 3 long bones: the **humerus**, **radius**, and **ulna**. The humerus is the single bone of the upper arm. The radius and ulna make up the forearm. As mentioned earlier with the scapula, the head of the humerus fits into the depression of the scapula called the **glenoid cavity** (glenoid fossa). At the opposite end of the humerus at the elbow, the humerus will form a joint with portions of the forearm bones.

The elbow joint is produced when the smooth surfaces of the humerus connect with the joining (articular) surfaces of the radius and ulna. The **capitulum** and **trochlea**, boney projections of the humerus, fit together with the head of the radius and the trochlear notch of the ulna, respectively.

At the ends of the forearms, a collection of bones cluster together into the wrists, hands, and fingers. The 8 small irregularly shaped bones adjacent to the radius and ulna are collectively called the **carpals**, and form a base upon which the longer bones of the hands and fingers are attached to the bones of the forearm. The first long bones that extend from the carpals to the fingers are the **metacarpals** and are attached to the 3 remaining bones of each finger, the **phalanges**.

The bones of the wrist and hands (carpals and metacarpals) sound much like the bones of the ankle and feet (tarsals and metatarsals). An easy way to remember the location of the carpals is to remember that you steer a CAR with your hands.

## Pelvic Girdle

Like many of the other skeletal structures of the human body, the pelvis is a composite structure composed of the previously described sacrum (along the dorsal aspect) and 2 hip bones, the **coxal**, or innominate, bones. Together these bones create a bottomless, basket-shaped girdle that supports the lower abdominal organs as well as providing an attachment point for the legs back to the axial skeleton. Additionally, this open-ended structure is wider in females than in males to facilitate the passage of a baby through the birth canal.

The 2 hip bones are joined to the sacrum along the dorsal aspect of the body at the **sacroiliac joint**. This is a relatively fixed joint, especially when compared to the joining of the 2 pelvic bones along the ventral aspect of the body at the **pubic symphysis**, the front portion of the hip. During pregnancy and at delivery, this joint allows for the expansion of the pelvic girdle and delivery of the baby through the birth canal.

Each pelvic bone has 3 regions. The broad blades on the back of the pelvis are the **ilium** portions of the hip bones. The ventral portion of the hip bone is divided into a superior (pubis) and inferior (ischium) around which is a large opening called the obturator foramen. Laterally on each hip bone at approximately the location where all 3 portions of the hip bone connect is a large depression where the ball of the femur (leg bone) connects. This depression is called the **acetabulum** and forms the socket of the hip ball and socket joint.

## Legs

Using the same basic layout as described in the arms, the legs are composed of a single upper leg bone (the **femur**), 2 lower leg bones (the **tibia** and **fibula**), and bones of the ankles and feet. The femur supports the full weight of the body. It joins with the pelvic bone via the head of the femur fitting into the socket of the acetabulum in the hip. At the knee, the **medial** and **lateral condyles**, boney projections of the femur, form the articular (joining) surface with the smooth surface of the tibia. This lower leg bone will support the full weight of the body whereas the fibula actually joins with the side of the tibia rather than directly contributing to the knee itself. The remaining bone of the knee, which is actually suspended within its own tendon, is the **patella** (kneecap) and seemingly floats on the surface of the knee.

At the ankle, the lower portion of the fibula (lateral malleolus) projects outward on the outer side. This rounded bump is clearly prominent on the surface. It is here, at the surface of both the tibia and fibula, where the ankle joint is formed between the bones of the lower leg and the **tarsal** bones of the foot. The most prominent of these tarsal bones is the **calcaneus** (heel bone). Named similarly to those in the wrists and hands, the first long slender bones attaching the toes (phalanges) to the foot are the **metatarsals**.

# Joints

While joints are often thought of as the moveable points of the human body, some are actually completely immoveable. Thus, the definition of a joint in strict anatomical terms is the connection between 2 or more opposing bones, which may or may not allow for movement.

## Ball and Socket Joints

This type of joint, as seen in the hip and the shoulder, is produced when the ball, or head, of a long bone inserts itself into the bowl-shaped depression of another bone, either of the pectoral girdle (glenoid cavity of the scapula) or the pelvic girdle (acetabulum of the pelvic bone). This type of joint allows for a wide range of rotational movement and gives the most flexibility of the body's joints.

### Synovial Joints

Synovial joints are also highly flexible, but only along a single plane. These are often thought of as "hinge" joints, much like a door may swing open or shut along a single plane. These joints (the elbow and knee) are covered in membranes that contain cells producing **synovium**, a lubricating fluid that can also cushion the surfaces of the bones that are in contact. Because the synovial fluid can withstand great pressures, even the pressure of gravity pulling the femur down onto the tibia and fibula will be unable to move the encapsulated fluid from the space between them. Thus, the fluid prevents bone-on-bone contact.

### Fibrous Joints

Fibrous joints are atypical joints in the sense that they are immoveable. In the adult human skull, these joints are seen as the suture lines between the bones of the skull (calvarial bones). During the fetal period and into early childhood, the bones of the skull are not attached to one another, to allow for the deformation of the head and thus facilitate passage of the head through the birth canal. However, with development and growth during the neonatal period (first 4 weeks of a child's life), these bones will contact each other and fuse permanently with only the suture line as evidence that the bones were ever separated at all.

## Bone Growth

Unlike the enamel of teeth, bone is living tissue, containing cells that can remodel the bone for normal growth and to repair any damage that may occur. During embryonic and fetal development, bones form in one of two ways.

### Intramembranous Osteogenesis

The flat bones of the skull, clavicle, and sternum are among the bones that form by a process known as **intramembranous osteogenesis**. Derived from the embryonic membranes, connective tissue cells that will become bone cells start to gather and divide in the future sites of the bones. These cells will form early bone cells called **osteoblasts**, which begin the process of forming the calcium phosphate–rich bone matrix. This process will continue until the bone cells completely surround themselves with bone matrix, at which time they become osteocytes and reside inside cavities in the bone called **lacunae**.

Since bone matrix is too dense for the diffusion of gases and nutrients to occur, osteocytes are interconnected with adjacent osteocytes via **cytoplasmic extensions** that extend through small tunnels in the bone matrix and form a network of canals (canaliculi). Bone cells pass materials along the chain of interconnected cells much like a bucket brigade passes buckets of water to a fire. Each cell along the way will use the materials it requires to survive and pass the remainder along the chain to the next cell.

In this way, these relatively flat bones will grow and form plates of bone that resemble a cream-filled cookie. There will be 2 outer layers of highly compacted bone called the **tables** with a region between of more spongy bone (**diploe**) and a marrow cavity.

## Endochondral Osteogenesis

Long bones, such as the humerus and femur, are first formed during development as small cartilaginous templates that exhibit the same rudimentary shape as that of the adult bone. As the fetus grows, this template will also grow until blood vessels penetrate the middle of the long shaft (**diaphysis**) of the bone. These vessels will bring in and deposit bone stem cells into the region of cartilage and these cells will begin to grow.

First a collar of bone will grow around the middle of the shaft and will form what is termed a **periosteal** bone collar. Since cartilage is fed by diffusion, this collar of dense bone will suffocate the cartilage cells, which will begin to die, leaving room for the newly deposited bone cells to start to fill in. This primary **ossification** (bone-forming) center begins the restructuring of the shaft of cartilage into bone. The outer portion of the shaft will be composed of layers of compact bone with very little spacing other than that which allows the passage of blood vessels and nerves along the length of the bone. In the center, the bone will form into shards called **trabeculae**, and leave much space and appear spongy. This will be the location of the marrow cavity, which will fill with fat (yellow marrow) or blood stem cells (red marrow).

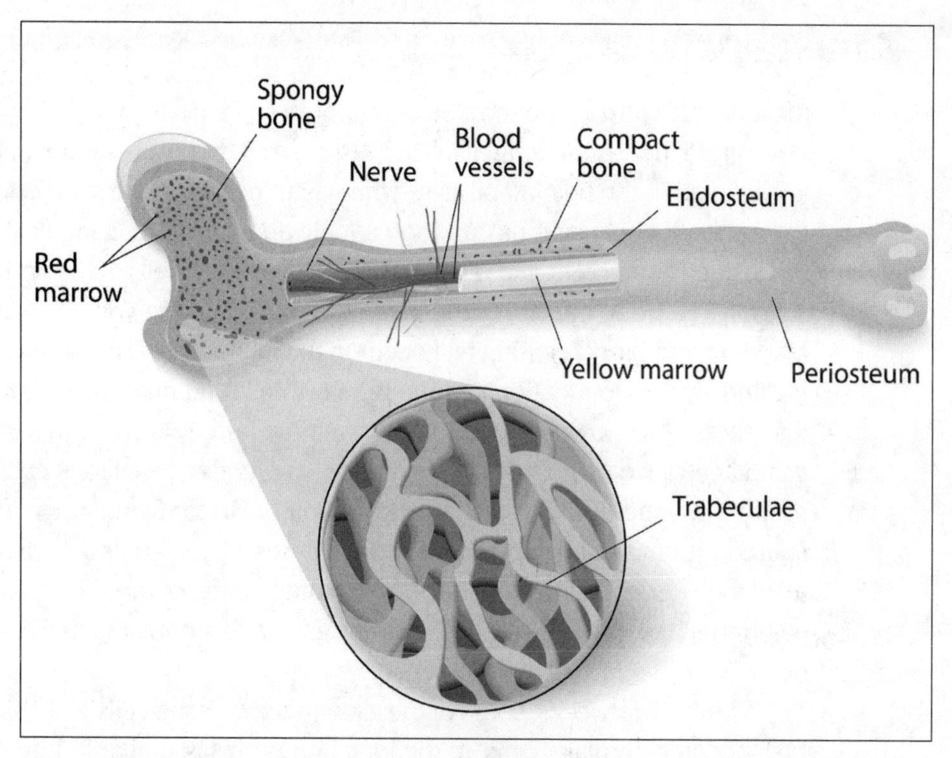


This illustration of a human femur shows the regions of the bone as well as the woven nature of the spongy bone next to the marrow cavity.

While these processes shape the shaft of the bone, vessels also enter the bulbous heads of the bones called the **epiphyses**. In a similar fashion, cartilage will die and be replaced by bone precursor cells and form a secondary ossification center. With bone forming from the middle of the shaft upward (and downward) and the secondary centers forming bone from the ends toward the middle, the only cartilage that will remain is what is caught in between these 2 ossification centers at the junction between the **epiphysis** and diaphysis. These plates of cartilage will persist until the individual is in his early to mid-twenties, at which time all cartilage will have been replaced with bone. During childhood and especially at puberty, these **epiphyseal plates** (growth plates) will add more cartilage cells on the epiphyseal side while the side adjacent to the shaft is replacing the cartilage with bone. Thus, with the addition of cartilage and the conversion to bone, this will lengthen the bone for continued growth.

## Bone Repair

Most people will experience a broken bone at some point in their life. The great news is that bone can repair itself in most cases. This process is not unlike what is observed in the healing process when a cut occurs in the skin. Bone is able to heal itself because it has a blood supply and is rich in living cells both on the inside (bone cells) and outside (connective tissue cells of the **periosteum**).

When a bone breaks, blood vessels within the bone will be severed, and bleeding and blood clotting will occur at the injured site. This stops the bleeding and lays the foundation for inflammatory cells and immune cells to migrate into the site to prevent infection and remove any pathogen that may have gained entrance to the body. Shortly afterward, connective tissue cells that surround the bone in the periosteum layer will begin to divide, migrate into the injured site, and start laying down connective tissue materials, which are equivalent to the granulation tissue seen in wound healing of the skin. This material will bridge the gap between the broken ends and stimulate changes that will occur at both edges of the broken bone.

Within a few weeks, some of these connective tissue cells will change and, first, become cartilage cells. In the long bones, this is similar to how the bone first forms during embryonic development, which is as a cartilage template. This time, the cartilage will be used to provide structural material that will fill in the gap between the broken ends where bone will be deposited. Osteoblasts then produce bone, which will grow into the newly formed cartilage and complete the repair process. These new layers of bone, while they will be intact, will be modified slowly over time, just as all bone is modified and reshaped as the body ages.

## Diseases and Disorders

As mentioned before, bones are rich in calcium, which is a key ion in many physiological activities in the body. Calcium can be deposited into bone and also recruited from the bone to be used elsewhere. Over time, the balance between give and take of calcium from the bones can become unbalanced and result in weakening of the bone. Additionally, if certain key minerals and vitamins aren't a part of the human diet, the process of normal bone growth can become unbalanced and lead to bone growing irregularly. In this section, we will look at two such problems related to bone growth.

## Osteoporosis

This degenerative bone disease occurs primarily in women of postmenopausal age and leads to a weakening of the bones in the body, sometimes resulting in bone fractures. Because the bone is losing mass or becoming more porous, the disease was named after this histological appearance. After menopause, hormone levels, especially those of estrogen, fall below what is needed to maintain the balance between bone absorption and bone deposition. Thus, cells called **osteoclasts**, which remove bone matrix and free calcium for return to the blood stream, become more active than the osteoblasts (bone-depositing cells).

Additionally, the hormones that control calcium reabsorption and storage also appear to change in favor of calcium restoration into the blood stream, which assists in the loss of bone mass with age. Hormone replacement is a clinical option for these patients (as is calcium supplementation); however, certain hormones have been shown to increase the risk of breast cancer and must be done under the close supervision of a physician.

## Rickets

Bones in individuals with rickets are typically seen as bowing in the long bones of the legs or otherwise misshaped bones within the arms or legs. The underlying cause of this disease is inadequate calcium deposition into bones, resulting in thinner and weaker bones (which bow under the weight of the body). While calcium may be present in the diet, what is missing is vitamin D. This is essential for proper absorption of calcium across the wall of the intestinal tract and transport into the blood stream where it can be utilized by the body. Likewise, calcium deficiency may also lead to rickets; however, this typically only occurs in areas where people, especially children, are living in conditions of famine and starvation.

CHAPTER 7

# Muscular System

Skeletal muscles are the engines that enable the human body to move and perform physical tasks as simple as holding a coffee cup or as complex as ballet dancing. Muscles provide the force while the bones act as the levers; the nervous system coordinates these actions. While there are 3 types of muscles in the human body (skeletal, cardiac, and smooth), only skeletal muscle will be described in the following sections.

# Major Skeletal Muscles

Depending on how you divide up the skeletal muscles in the human body, there are well over 600 named muscles, with some estimates reaching as high as 800-plus. In the sections to follow, many of the superficial and more well-known and recognized muscles will be described along with the actions they mediate.

## Muscles of the Head and Neck

The lips of the human body, which can be so very expressive, are in fact a single oval-shaped muscle called the **orbicularis oris**. Likewise, surrounding each eye is a circular muscle called the **orbicularis oculi**, which is responsible for closing the eyelids. There are many other smaller muscles for facial expressions, including the **frontalis**, the forehead muscle that raises the eyebrows, and the **buccinators**, which form the cheeks and pull them inward when contracted. For chewing, no other muscle is as important as the **masseter**. Its insertion into the mandible leads to the powerful closing of the jaw during eating.

Several muscles of the neck help to rotate and turn the head from side to side and up and down. The most prominent of these muscles is the **sternocleidomastoid**. It runs from the side of the skull around to the ventral side of the lower neck and connects with the sternum, which leads to the action of pulling the head to the side of the contracted muscle. Since this is a paired muscle, with one on the right and one on the left, together they give the appearance of a V-neckline on the ventral lower neck. On the dorsal surface, a portion of a larger muscle, the **trapezius**, connects to the back of the skull as well as to the cervical and thoracic vertebrae in a fanlike pattern and inserts into the scapula. This arrangement leads to various movements of the scapula depending on which portions of the muscle are contracted.

## Muscles of the Chest and Shoulders

The muscle that forms the bulk of the human chest is the **pectoralis major**, often simply referred to as the pecs. Much like muscles along the back, the pectoralis attaches to the sternum in a fan-shaped pattern, with the fibers coming to a point and inserting into the upper portion of the humerus of the forearm. Contraction of this muscle causes the humerus to be pulled toward the ventral midline of the body, therefore adducting (moving) the arm.

An adduction is a muscular action in which a body part is moved toward the midline of the body. You can remember this by saying the movement ADDs to the midline. If the body part is moved away from the midline, the action is referred to as an abduction.

Abduction of the arm is largely accomplished by the action of the **deltoid** muscle. This large muscle makes up the bulk of the shoulder. It has a dorsal, a ventral, and a medial portion, all attached to both the scapula and the clavicle. With an insertion on the humerus, contraction of this muscle will lead to the arms being raised (abducted) much like a child pretending to fly. While the pectoralis and the deltoids are the most prominent of the muscles that move the arm, there are many other smaller and deeper muscles that will mediate all other actions of the arm and allow it a wide range of motion and action.

## Muscles of the Arms

In the upper arm there are 2 muscles that work rather antagonistically, or against each other, in flexing (bending) or extending (straightening) the forearm. The large muscle on the ventral surface is called the **biceps brachii**, which is incompletely divided into a long and short head, hence the terminology (*bi-* meaning 2). It is the biceps that will pull the forearm closer to the upper arm when contracted and thus flex the arm. Pulling against the forearm and straightening it is the task of the **triceps brachii** on the ventral side of the upper arm. The name of this muscle stems from the division of this muscle into 3 heads, 2 of which are visible superficially and the third being deep.

The forearm is a highly intricate collection of small muscles that will assist in moving the forearm as well as moving the wrists and fingers. The muscles only compose the most **proximal** (closest to the body) portion of the forearm and connect to bones of the wrists and fingers via long tendons, each of which will lead to a specific action. A number of these are referred to as **flexors** and **extensors** because of the actions they facilitate. Additionally, other muscles will mediate **pronation** (twisting of the hands from a palms-upward position to a palms-down position). Conversely, other muscles called **supinators** will rotate the hands back to a palms-up position.

## Muscles of the Back and Hips

Starting at the highest aspect of the back (dorsal surface) is the aforementioned trapezius muscle. The remainder of the superficial back muscle is the large fan-shaped **latissimus dorsi**. Attached along the vertebral column from the thoracic region downward to the sacrum, it extends to a point under the arm and inserts into the humerus. In bodybuilders, contraction of this muscle causes the edge of the muscle to extend our laterally on both sides of the body, much like the hood of a cobra fanning out around the head.

In addition to adducting the arm, the latissimus dorsi will also pull the shoulder downward. This is one of the main muscles used when someone performs the chin-up exercise. Other, deeper muscles of the back include the rhomboids (major and minor), which also attach to the vertebral column and cause various movements of the scapula when contracted.

Making up the dorsal hips, or what many refer to as the posterior, are the 3 **gluteus** muscles. The gluteus maximus is the largest of the 3 and produces the bulk of the posterior hip tissue. When contracted, this muscle will extend the hip and bring the thigh into a straight line with the hip. Along the side of the hip near the waist is the **gluteus medius**, which along with the deep gluteus minimus will abduct and rotate the thigh to the side.

## Muscles of the Abdomen

While not everyone has a clearly visible six-pack, everyone does possess the paired rows of 5-muscle bundles that lie along the ventral midline of the body, which are referred to as the abdominal muscles, or **rectus abdominis**. It is the middle 3-muscle bundles that are often called the "6-pack." This muscle, as well as the following ones described, is responsible for tensing the abdominal wall and compression of the abdominal contents. From the rectus abdominis and running along the sides of the body are the external **oblique** muscles. The rectus abdominis and the external obliques are the superficial muscles of the abdominal wall.

Just beneath the external oblique is the **internal oblique**. The **transversus abdominis**, the deepest of the abdominal muscles, lies underneath the internal oblique muscle.

**How many muscles actually make up the abdominal muscles visible in muscular individuals?**
While many people work out to achieve the "six-pack" ab look, there are actually eight rows of abdominal muscles that make up the ventral surface of the abdomen. The lower pair are longer than the six-pack and are the most difficult to make visible.

## Muscles of the Legs

On the ventral surface of the thigh, there is a major group of muscles collectively referred to as the **quadriceps** (quadriceps femoris). This group comprises the rectus femoris, vastus lateralis, vastus medialis, and vastus intermedius. The prominent rectus femoris is the middle and superficial muscle of the group, with the vastus lateralis and medialis on the outside and inside of the thigh, respectively. The intermedius is a deep muscle lying underneath the rectus femoris. Together, these muscles function to extend the knee (straighten the leg). Standing from a squatting position would be the work of the quadriceps.

On the back of the thigh (dorsal side), the group of muscles that function to flex the knee and pull the heel upward toward the hip is called the **hamstring** group. The 2 superficial muscles of this group are the **biceps femoris** and the **semitendinosus**. These paired muscles lie along the midline of the dorsal thigh with a deeper, more medial (inside) muscle called the **semimembranosus** completing the group.

The calf muscle is formed from the 2-headed **gastrocnemius** muscle. This muscle is connected to the heel (calcaneus bone) via the long **calcaneal tendon** (also known as the Achilles tendon). Contraction of this muscle will pull the heel upward and function to straighten the foot. This action can be exemplified as someone standing flat-footed and then rising up on the tips of her toes, much like a ballet dancer. Additionally, as seen in the hand, many smaller muscles will function to flex, extend, and rotate the foot in several different directions.

# Neuromuscular Junction

Muscles are instructed to contract by the nervous system and the electrical signals that flow from neuron to neuron. However, there is no physical contact between neurons and target tissues, such as muscle. Therefore, the electrical signal of neurons must be changed into a chemical signal that can diffuse across the synaptic cleft between the neuron membrane and that of the muscle. The chemical receptors on the muscle cell detect the chemicals and activate a signaling cascade that will lead to the regeneration of the electrical signal within the muscle cell. This causes the contraction. In the following sections each component of this connection (called a synapse) and signal transduction mechanisms will be discussed.

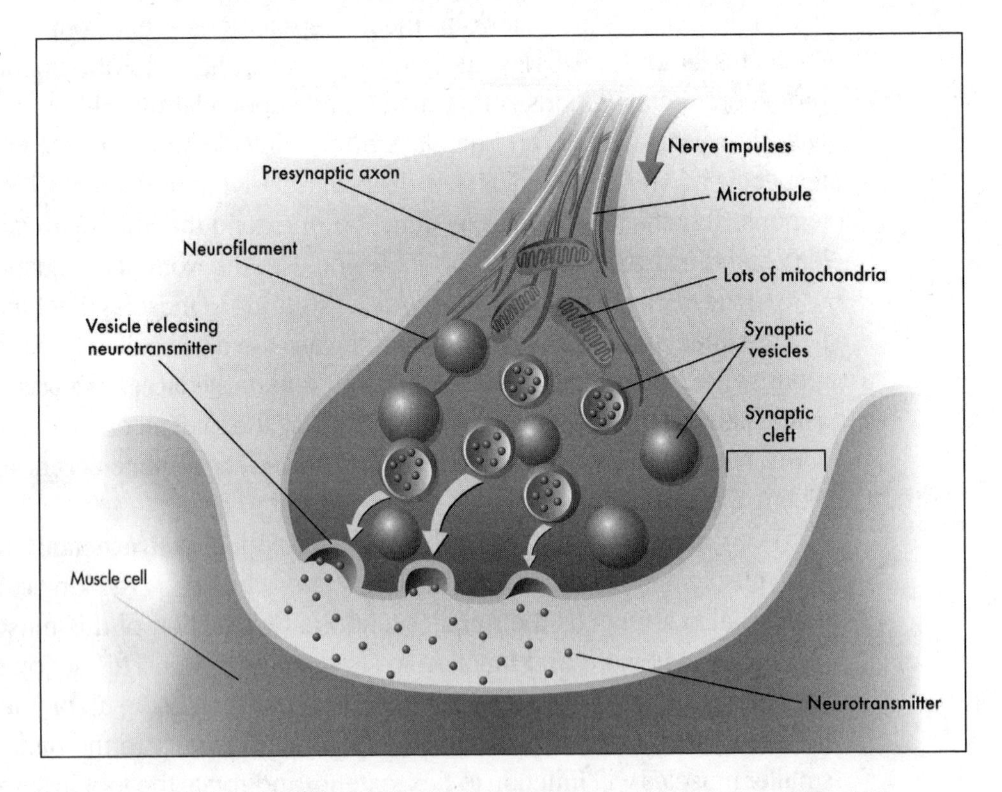


Nerve impulses lead to the secretion of neurotransmitters (chemical signals), which are detected by receptors on the muscle cell surface and result in a muscle contraction.

## Motor Nerve

Signals from the nerves result in the contraction of skeletal muscles. In nerve cells (neurons), differences in charged particles (ions) between the inside and the outside of the cells result in a membrane voltage, with one side of the plasma membrane being more or less positive than the other. Think of this as being much like the battery in your remote control with its positive and negative ends (poles).

As the neuron receives a stimulus, membrane channels allow ions to move across the membrane and cause the voltage of the membrane to change. This localized change of the membrane will influence neighboring membrane channels, and also opens channels in a wavelike progression that flows along the axon (the long process of the neuron) toward its target. When this moving change of voltage (action potential) reaches the end of the nerve, chemicals stored in vesicles fuse with the terminal (presynaptic) membrane and are released into the space between the neuron membrane and the target cell. The chemicals that are released from neurons and bind to receptors on target cells are collectively called neurotransmitters. Just as perfume diffuses through the air, is detected by an individual, and causes a pleasant sensation, these chemicals diffuse from one cell to the target cell and result in an action. For motor neurons that stimulate skeletal muscle contraction, the neurotransmitter secreted is **acetylcholine**. Thus, the electrical signal of the nerve has been transduced into a chemical signal that can diffuse across space.

## Skeletal Muscle Membrane

To regenerate the electrical signal in the muscle cells, neurotransmitter receptors are localized in close proximity to the neuron terminal on the muscle membrane. In skeletal muscle, these acetylcholine receptors are named **nicotinic receptors**. When 2 molecules of acetylcholine bind to the receptor, sodium and potassium ions are allowed to flow into and out of the cell through the nicotinic receptor. This localized change in voltage also leads to changes in the membrane farther away along the muscle surface. These regions have membrane channels like those along neurons, which facilitate a wavelike action. Once activated by the nicotinic receptors, the action potential is regenerated and flows over all of the muscle membrane. Because the plasma membrane of skeletal muscles projects downward into the muscle cell, forming tunnels called **T-tubules**, the action potential spreads rapidly and completely throughout the muscle cell and results in a muscle contraction.

### Motor Units

For muscles to contract, energy (in the form of ATP, adenosine triphosphate) will be expended. To conserve energy and use only as many muscle cells as are required to accomplish a task, muscles are divided up into functional units called motor units, which consist of a single motor neuron and all of the muscle cells that the neuron is connected to and controls. To initiate a contraction, the central nervous system will activate only a few motor units and then progressively activate more and more until the work is accomplished. For a muscle whose primary function is strength, (e.g., quadriceps group), there may be several hundred muscle cells connected to a single motor neuron. However, for muscles that require less strength and more control, such as the ciliary muscle of the eye that attaches to and controls the shape of the lens, only a few muscle cells are attached per neuron.

## Muscle Contraction

Muscles have only one function, and that is to contract. For skeletal muscle, this results in a shortening of the muscle in a single plane, pulling both ends of the muscle (and the structures to which they are attached) closer together. This is mediated by the 2 contractile proteins, actin and myosin, and the sliding movement that occurs between them during a contraction.

### Banding Pattern of Striated Muscle

Skeletal muscle (and cardiac muscle) is often described as striated muscle because of the striped appearance of the individual muscles cells when observed with a microscope. These alternating light and dark stripes or bands form the foundation and the basic unit of striated muscle organization and are essential for these cells to contract.

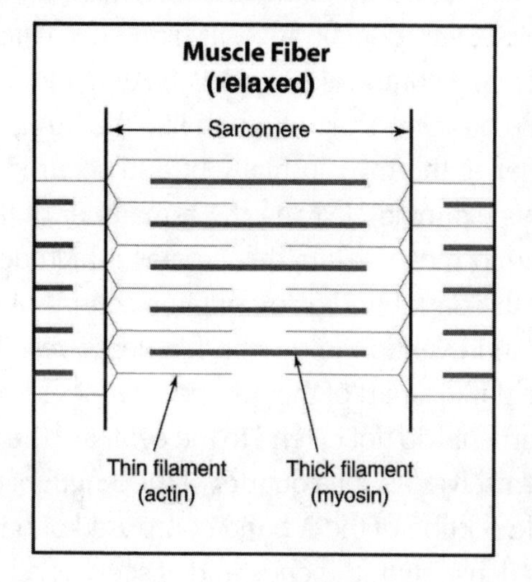


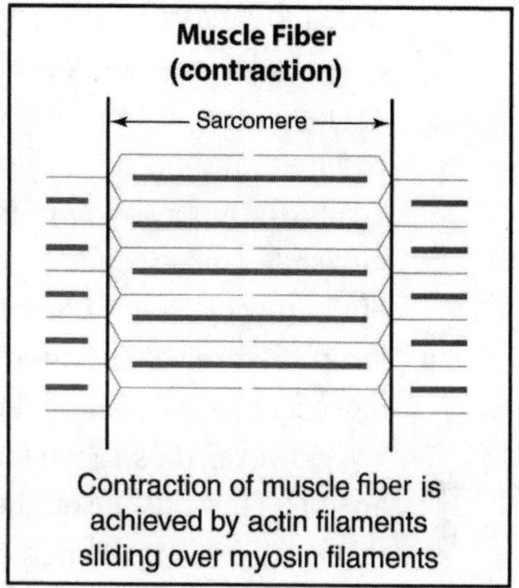


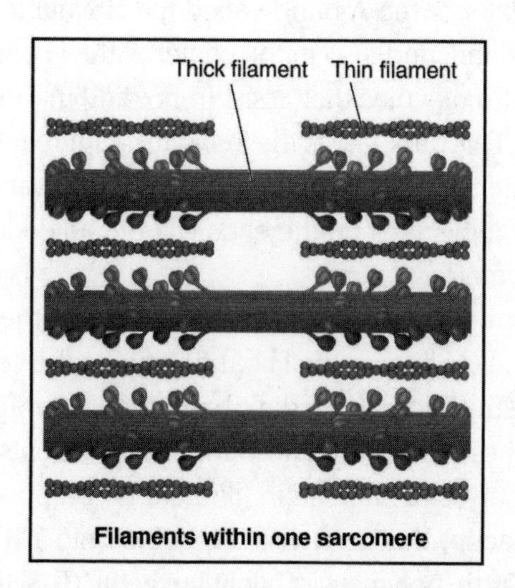


**The overlapping actin and myosin contractile proteins are arranged in a specific order that results in a repeating banding pattern in muscle cells. This repeating unit of muscle is called a sarcomere. The actin and myosin move closer together during a muscle contraction.**

The light and dark bands are caused by the amount of light that may pass through a particular region of the muscle. The denser areas will be darker, leaving the less dense regions lighter by comparison. These darker bands, called **A bands** or **anisotropic bands**, contain myosin filaments. Often referred to as thick filaments, it is actually composed of between 200 and 300 individual myosin molecules. These proteins, shaped much like the human arm, are wrapped around the cylinder of the myosin filament and secured by a light chain (the upper arm, in this example), leaving the remainder of the myosin molecule (heavy chain) free to move. While this creates a high-density area of the muscle, each end of the dark band also contains actin molecules that insert themselves between and overlap the myosin filaments, resulting in the highest-density area and the darkest part of the A band.

However, these actin filaments do not extend to the center of the myosin filaments. They only insert themselves about a quarter of the length of myosin on each end, leaving the middle portion of the A band composed of only myosin. This region is therefore less dense than the ends and is seen as a light region in the center of the A band called the **H band**. Additionally, in the very center of the A band, and also in the center of the H band, is a dark line consisting of structural molecules that assist in holding the myosin filaments in the proper position. This dark line is the **M band (M line)** and also marks the center of the contractile unit of skeletal muscle called the **sarcomere**. During a contraction, the actin molecules on the ends will be pulled toward the M line and shorten the sarcomere.

In summary, the dark A band will contain the light H band in its center with a dark line (M line) in the H band defining the very center.

On either side of the dark band are lighter regions called the **isotropic (I) bands**. These areas contain only actin filaments, which are much less dense than the myosin. Each actin filament is composed of 2 strands of **filamentous actin (F-actin)** that are twisted together into 1 filament. Each F-actin strand is formed from polymers of **globular actin (G-actin)** molecules. This gives the F-actin the appearance of a pearl necklace, in which each pearl represents a G-actin molecule. The I band is only interrupted by a dark line (**Z line** or **Z disk**) that defines the center of the I band and is composed of structural molecules much like those of the M line in the dark band. These molecules also assist in maintaining the proper spacing of the actin molecules, which is critical to the sliding filament action that will occur during a muscle contraction.

As mentioned earlier, the contractile and repeating unit of striated muscle is called the sarcomere and is defined as the filamentous region between

2 adjacent Z lines. This will contain an entire A band and 2 halves of I bands on each end. Thus, during a muscle contraction, as actin molecules are pulled closer to the M line, the Z lines will be pulled closer together and the sarcomere as a whole will shorten.

## Accessory Proteins

While the structural molecules of the M and Z lines are important for the alignment of the thin actin and thick myosin filaments, other molecules play essential roles in helping regulate a muscle contraction and in returning the muscle to the relaxed state. One such molecule is **tropomyosin**. Two of these filamentous molecules will run along the grooves between the 2 F-actin molecules (which compose a thin actin filament) and function to cover or mask the sites on each G-actin molecule where myosin can bind. When tropomyosin is in this position, the muscle is relaxed.

Attached to tropomyosin is a multiunit molecule called **troponin**. One of troponin's 3 subunits, troponin I, binds to a region of the actin molecule. The troponin T subunit binds to the tropomyosin molecule, and the troponin C subunit is capable of binding to a calcium ion. Calcium is the key to unmasking the myosin bind regions of the actin molecules, resulting in a muscle contraction.

Although named tropomyosin, this molecule is actually wound around F-actin strands, therefore blocking the binding sites for myosin binding to actin. The naming of this molecule often confuses students on its location and function, so commit this to memory.

## Calcium and Its Role

In the relaxed state, calcium is stored in muscle cells inside organelles called **sarcoplasmic reticula**. This calcium reservoir is responsible for releasing calcium upon nerve stimulation, and also for pumping calcium back inside when the nerve signal ceases, signaling the muscle to relax.

Voltage-gated calcium-release channels are closely associated with the T-tubules described earlier. As the action potential spreads across the surface and down into the T-tubules, it will lead to the rapid release of calcium from the sarcoplasmic reticulum and its rapid spread throughout the cytoplasm of the muscle cell, where it encounters the troponin C subunit of troponin.

Each myosin molecule can be illustrated by using your arm. Think of your arm, from shoulder to hand, as a myosin molecule with two hinge points (elbow and wrist). From shoulder to elbow would be analogous to the light chain (light meromyosin) of myosin, and from elbow to (and including) the hand is the heavy chain (heavy meromyosin). The hand in this analogy is the S1 region (domain) and is the portion that will bind ATP and release energy for the contraction, as well as connect with a G-actin molecule to enable the sliding of filaments.

When calcium binds to troponin, the entire molecule will change shape. With the attachments to tropomyosin and actin, tropomyosin pulls away from actin and exposes the myosin binding sites on the G-actin molecules. Once free, the head of each myosin molecule (the S1 portion of heavy myosin) connects with G-actin and begins a muscle contraction.

## Sliding Filament Motion

When the head of myosin connects with a G-actin, a cross bridge is formed. This is made possible because prior to the connection, a molecule of ATP was bound inside the myosin head and split into separate molecules of ADP (adenosine diphosphate) and phosphate, which remain inside. Once the cross bridge is formed, however, the phosphate molecule is released, which triggers a change in the myosin molecule called a **power stroke**. Since the myosin head is bound to actin, the power stroke is the action that results in the sliding of actin closer to the M line. Since this is happening on either side of the M line, each Z line is moved closer together and the muscle as a whole contracts.

This single contraction cycle only provides a fraction of the shortening distance for the muscles. Several repeated cycles of contraction will need to be accomplished for the entire muscle to shorten the full distance. Thus, each myosin head will need to undergo a power stroke, release, reset, and repeat a number of times to shorten the muscle fully. Consider a tug-of-war team. Each individual pulls on the rope, trying to pull the other team across a line. During the contest, it will be necessary for members to release their grip and pull on the rope from a new position. If each member released at the same time, the team would lose. So the members will release and form new grips in an alternating fashion. Such is the case for cross bridges for striated muscle.

Rigor mortis, or the "stiffness of death," occurs as skeletal muscle cells contract at the time of death from an explosion of electrical signaling throughout the body. This signal is unable to cease and the muscles remain locked in continuing cycles of contraction. However, since ATP is required for contraction and for resetting, this will only occur until the supply of ATP is exhausted, at which time the muscles are locked in the last contracted power stroke and unable to rest.

The muscle release after a power stroke occurs as the remaining ADP molecule is displaced by a new ATP molecule in the myosin head. Therefore ATP is needed to contract and relax the muscle. During this release and resetting period, the ATP is hydrolyzed into ADP and phosphate, and as long as the myosin binding site is still exposed on actin, a new cross bridge and power stroke can occur.

At the end of a muscle contraction, the nerve stops sending neurotransmitters and stops the action potential on the muscle cell. This halts the release of calcium and initiates the active transporting of calcium back into the sarcoplasmic reticulum. With calcium no longer available to bind to troponin, it returns to its original shape and allows tropomyosin to slide back into its original position. This blocks the binding sites on actin and leaves myosin in its resting position.

## Energy Sources for Contraction

After reading the previous section on muscle contraction, it is clear that energy, in the form of ATP, is required for muscle to shorten. In the human body, energy is stored in several forms and in different compartments.

At the onset of muscle contraction, the raw material that will be used to produce new ATP is pulled from the plasma of the blood and becomes circulating glucose. This carbohydrate can be plugged directly into the aerobic respiratory metabolic pathway for the immediate production of energy. As the work continues or intensifies and plasma glucose is in short supply, muscle cells will draw on plasma triglyceride resources for energy production. Materials available in the plasma are typically provided by the diet and are present for rapid use.

If the demand on muscles continues and all plasma resources diminish, the body will recruit energy stores first from triglyceride stores, such as those present in fat cells. Finally, for high-intensity work, the demand for energy by muscle cells exceeds what can be supplied by plasma molecules or fat cells. In this case, the body recruits glucose from storage. The liver is one organ adept at storing glucose as **glycogen**, which can be tapped to return glucose into the plasma, supplying the muscles with a rich source of raw material for immediate ATP production.

## Diseases and Disorders

Problems with the muscular system can range from mild and temporary discomfort to the extreme of muscle loss (atrophy) and death. Most will experience the first at some point, while the extreme conditions are much less common. However, it is likely that you will encounter someone who has been affected by muscular disease in your lifetime.

### Muscle Spasms

Resulting largely because of dehydration, muscle spasms are involuntary and repetitive or constant contraction of a muscle without being able to relax. Often uncomfortable and painful, these spasmodic contractions occur because the muscle has received an inappropriate electrical signal to contract and remain contracted until the electrical signal stops. Recall that muscles are signaled to contract by a nerve stimulating a change in voltage on the muscle membrane. If the concentration of these charged ions changes as a result of dehydration and become more concentrated around the muscle cell, this can yield the same change in voltage as a nervous stimulus and lead to a contraction. Since this is not regulated as a normal contraction, there is no halt to the signal until the ions becomes balanced once again. Often, therapists or trainers will massage and stretch the affected muscle in an effort to increase the circulation in and around the muscle in order to balance the ions as quickly as possible. Additionally, immediate rehydration will aid in preventing the recurrence of the cramps.

## Muscular Dystrophy

While many forms of muscular dystrophy have been described, the most common form is **Duchenne muscular dystrophy** (DMD), which occurs most commonly in childhood. This debilitating disease not only reduces muscle mass and mobility, it also shortens the life expectancy of the patients considerably.

The cause of DMD was determined to be a mutation of the gene for **dystrophin**, a protein that interconnects the muscle cytoskeleton to the extracellular environment through the muscle membrane. Loss of dystrophin function results in highly disorganized muscle, smaller muscle mass, and an increase in connective and inflammatory tissue.

**Becker muscular dystrophy** is a variant of DMD where dystrophin is shortened but remains functional, and is therefore a less severe form of Duchenne muscular dystrophy.

CHAPTER 8

# Nervous System

The nervous system is one of the major control centers of the body. It is composed of sensory receptors and nerve tracts, which bring information about the body (internal and external) to the brain. The brain then processes the information and determines how to respond. Those responses leave the brain via different nerve tracts, travel through the spinal cord, and are dispersed through nerves to the appropriate target, such as different tissues. All of this information is moved around the body through ions and chemicals in and around the nerve cell. Therefore, the signal transduction of information is essential for the nervous system.

## Signal Transduction

As discussed earlier, neurons use their plasma membrane to selectively concentrate ions in either the cytoplasm or outside of the cell, which will affect the voltage of the membrane. This is not unlike a battery that will have a positive pole and a negative pole, which are separated from each other. For cells, that separation is the membrane, and the concentrations and charges of ions will result in an electrical polarization of the membrane.

When cells are inactive (at rest), the metabolic machinery of the cell will work to create a concentration gradient (a measurement of how the concentration of something changes from one place to another) of 2 critical ions: sodium and potassium. A membrane protein called the sodium-potassium pump ($Na^+$/$K^+$ pump) will use energy to move sodium outside of the cell while at the same time moving potassium inside. This, as well as the number of fixed charged particles on the inside of the cell (such as DNA, charged proteins of the cytoskeleton, etc.), results in the interior of the cell being more positive than the outside of the cell. This imbalance results in a membrane voltage of approximately -70 mV (millivolts). Since the cell is at rest, this voltage is termed the **resting membrane potential**.

The use of the term "potential" is significant. The cell will move these ions into the separate compartments, creating high concentrations of ions on opposite sides of the membrane. This is not unlike water behind a hydroelectric dam that can be used to create electricity as the water flows downhill through turbines. In the same way, the "potential energy" of these concentrations can be used to allow the ions to flow "downhill" through membrane channels and affect the voltage of the cell (which is the electrical signal).

## Voltage-Gated Receptors

When a stimulus imposes on a cell, protein receptors will cause some ions to diffuse out of their compartment and alter the membrane voltage. If this change is of sufficient magnitude, some membrane proteins will actually change their shape based on the electrical charge of the membrane. These are said to be "voltage-gated" channels. For instance, one such channel will change shape at a voltage of -55 mV in such a way as to allow sodium ions to follow their concentration gradient and diffuse into the cell. Since sodium is positively charged, this will cause the membrane voltage to become more positive and possibly affect other voltage-gated channels that respond at different voltages. Potassium voltage-gated channels open when the membrane potential is in the positive range. Think of this exchange of sodium and potassium much like two children playing on a seesaw. When one goes up, the other goes down, and vice versa. In this case, sodium floods in to depolarize the cell while potassium rushes out to repolarize the cell.

## Action Potential

As channels open and close and as ions change places, this will create a wave of voltage changes that have the capability to affect nearby areas of the membrane, causing them to undergo the same changes. This "wave" of voltage changes will move along the axon of a nerve cell in much the same way that a wave moves in the stands at a football game. In the stadium, as a person stands up then sits, the next person knows to then stand and quickly sit, and so on. On the cell membrane, as the sodium channels open at -55 mV and the movement of sodium causes the membrane potential to become more positive (and move toward 0 mV), this change is called **depolarization** of the membrane. Because sodium flows into the cell more rapidly than needed, the sodium channel will inactivate quickly after it opens to prevent too much sodium from coming in. However, this will still allow enough sodium in to change the membrane potential into the +30 mV range. Inactivation is different than the channel being closed, which occurs more slowly. This means that although the shape of the channel is open, another part of the protein has moved into place to block the passage of any further ions until the entire protein can reset and close.

As sodium rushes into the cell, it will disperse throughout the cell and directly under the membrane, leading to an increase of the membrane potential in the area of neighboring sodium voltage-gated channels, which are still closed. This will then lead to the opening of neighboring sodium channels if the membrane potential in that area also reaches -55 mV (in the same way the person sitting by you at the football game continues to do the wave). This will continue the length of the axon as long as there are sodium voltage-gated channels to be opened.

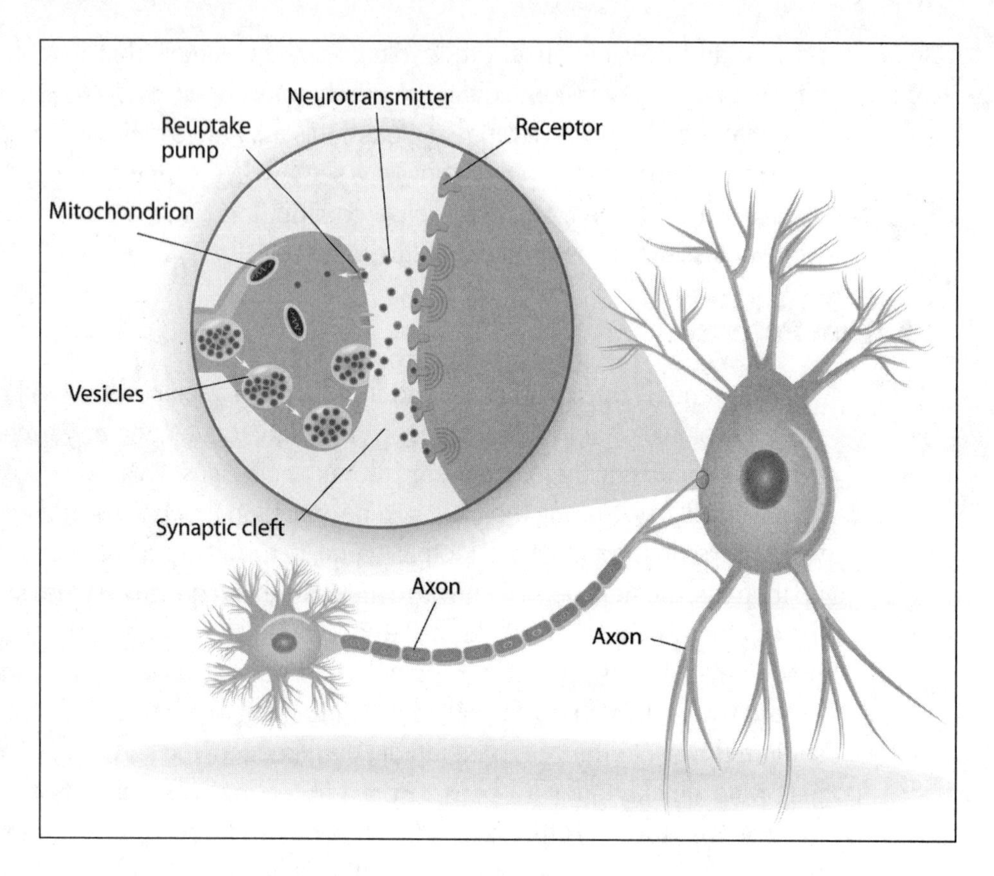


A synapse is a connection between two neurons. The neuron on the left is sending a signal to the neuron on the right by releasing neurotransmitters that the right neuron detects.

When the membrane potential enters the positive realm, potassium voltage-gated channels will open and potassium will rush out of the cell by diffusion. Since potassium is also positive, this will cause the potential to become more negative, therefore repolarizing the membrane. As with the sodium channels, the potassium channels also open quickly and inactivate suddenly, but not fast enough to prevent the potential from becoming even more negative than the resting membrane potential of -70 mV. This period is termed **hyperpolarization**, and although the voltage has returned to that near resting potential, the concentration of sodium and potassium ions has been reduced. To reset the ion concentrations, the $Na^+/K^+$ pump shifts into high gear to pump sodium out and potassium in, to recreate the high concentration gradients sufficient to have another action potential occur in the area of the cell.

It is estimated that the body's energy expenditure for the $Na^+/K^+$ pumps in the body is approximately 12 percent. This significant energy usage illustrates the importance of this enzyme in the balance and functioning of the human body.

## Neurotransmitters

The electrical wave that is the action potential works very well to transmit electrical signals at the speed of 1 meter per second. However, this will occur within a connecting cell. These electrical signals cannot move across space and continue in an adjacent cell spontaneously. Thus, to functionally interconnect neurons with each other and to their target tissues, the electrical signals are transduced into chemical messengers that can be secreted, diffuse through space, and be detected by the receiving cell that can transduce the now chemical signal back into an electrical action potential. In short, the electric signals are converted into chemical signals, which are able to move between cells.

## Types of Neurotransmitters

Neurotransmitters are the chemical messengers of the nervous system. There are several different types of these molecules that function with different tissues, are secreted by specific neurons, and elicit prescribed effects on the target tissues.

**Acetylcholine** (ACh) is the neurotransmitter most commonly secreted from motor neurons (nerve cells sending information out to the body) that cause skeletal muscles to contract. This is also the neurotransmitter used to stimulate portions of the involuntary nervous system. The result of the interaction with ACh and its receptor will always be an increase in the membrane potential of the receiving cell. Therefore, ACh is said to always be excitatory in nature.

**Norepinephrine** is another commonly used neurotransmitter that will affect smooth and cardiac muscle as well as glands of the body. Norepinephrine belongs to a group of neurotransmitters called **catecholamines** (molecules derived from the amino acid tyrosine) and functions largely in the involuntary nervous system to control those body functions either during rest or in fight-or-flight mode.

Other neurotransmitters, such as **dopamine**, **serotonin**, and **GABA** (gamma-aminobutyric acid), are found in the brain and control such activities as hunger, behavior, mood, and overall brain activity.

## Chemically Gated Receptors

Regardless of the neurotransmitter type, without a receptor to detect and instruct the cell, the chemicals would have no effect. Thus, the receptors are equally as important as the neurotransmitters themselves.

On skeletal muscle cells, the ACh receptor (nicotinic ACh receptor) will bind to 2 molecules of ACh that will result in a changing of the shape of the receptor. This will open a channel in the protein through which sodium and potassium can diffuse along their concentration gradients and result in a positive change in the membrane potential. Since these are chemically gated responses, it is not referred to as a depolarization; rather, this positive change is called an **excitatory potential**.

For smooth and cardiac muscles, there is a different ACh receptor: the **muscarinic** type. The binding of a single ACh molecule will lead to a shape change in this protein, but it is not an ion channel as you saw in the nicotinic receptor. This shape change will lead to a signaling cascade of molecules, much like a series of dominoes knocking each other over, that will end in either the opening or closing of ion channels. Depending on the ion and the directional effect, this type of signaling is either excitatory or **inhibitory** (increases or decreases the membrane potential, respectively). Receptors for other neurotransmitters in other tissues will similarly transduce signals, leading to alterations in their membrane potentials and yielding an appropriate physiological response.

## Synapse

A synapse is the connection between two neurons where the signal is relayed from one to the other via neurotransmitters. When the end of a nerve cell (the axon terminal) approaches the target tissue, such as a skeletal muscle cell, a space is created between the membranes of each cell (the synaptic cleft). It is across this membrane that neurotransmitters will diffuse when released from the axon terminal. Being on the receiving end of the chemical messages, the skeletal muscle cell membrane and the nicotinic ACh receptors that are present are said to be part of the **postsynaptic** cell or membrane. As previously described, the positive change in membrane potential, mediated by the nicotinic receptor, will allow ions to disperse throughout the cell and affect voltage-gated sodium channels that are nearby. Once the local membrane potential is -55 mV, the voltage-gated channels will open and the action potential will be regenerated in the new cell using the same machinery as previously described.

## Brain

The organ most responsible for the body's functions, conscious, and subconscious is the brain. The major component of the central nervous system (CNS), the brain is composed of the **cerebrum**, **cerebellum**, and the **brain stem**. All activities of the body are controlled from the brain, whether sensing the internal and external world, reflexively or consciously responding to stimuli, coordinating all body movements, or simply dreaming of a better world.

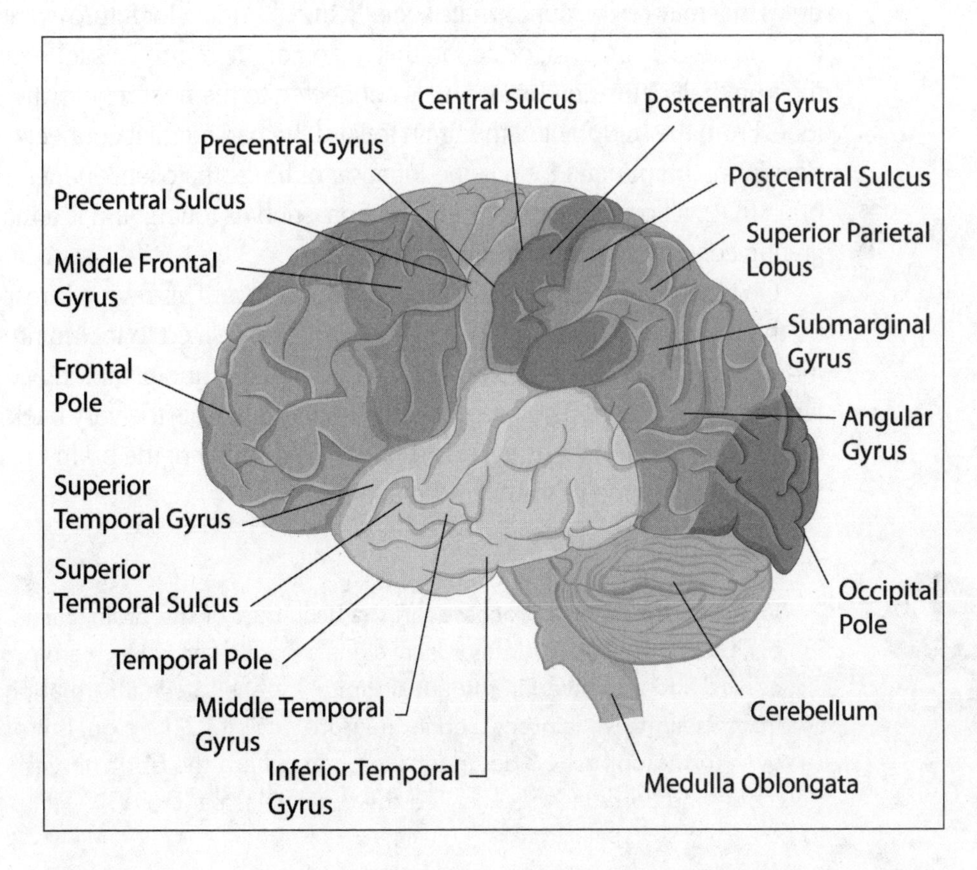


The human brain is composed of highly folded regions that are organized and specified for their function.

## Cerebrum

The cerebrum, or the forebrain, is expanded in humans and is what sets us apart from all other organisms on the planet. It is the organ used for conscious thought, for receiving and perceiving sensory information, and for initiating motor responses. On the surface, there are numerous folds (**gyri**) and grooves (**sulci**) that aid in the increased surface area for the outer portion of the brain, called the **cortex** of the cerebrum, to be packed into a smaller space. Thus the cortex of the brain will contain the nerve cell bodies of the brain cells and will exhibit a dark color compared to the deeper region. This is why the cortex is referred to as the **gray matter** of the brain. The deeper portion of the brain through which the myelinated axons run is called the **white matter** because of its translucent color.

Each hemisphere of the cerebrum is functionally divided into regions. The **frontal** region, or lobe, is involved in decision making, overriding instinctual urges that may not be appropriate socially, and planning for future events. Long-term memories are also formed in this region of the brain. This lobe makes up the front half of the cerebrum and is connected to the next region, the **parietal** lobe. From the midpoint of the brain toward the back until it connects with the next lobe, the parietal lobe is the location of the somatosensory center of the brain that will process sensory information, such as touch, and is a major integrating center for incoming information.

On the side of the cerebrum is the **temporal** lobe. This region is responsible for the processing of vast amounts of information related to incoming sensory information. New memory formation as well as the understanding of spoken language occur here. The final lobe, the **occipital** lobe, is the very back portion of the brain and functions as the primary visual center of the brain to interpret, integrate, and perceive visual information.

Visual information is processed in the back part of the brain, called the occipital lobe. Interestingly, visual signals from the right eye are processed and perceived in the left occipital lobe, and eye information from the left eye is processed in the right occipital. The signals cross over to the opposite side under the brain, where the optic nerves meet, and cross in what is called the optic chiasma (cross).

### Cerebellum

Positioned beneath the occipital lobes of the cerebrum, the cerebellum consists of 2 major lobes, called cerebral hemispheres, with a smaller interconnecting region, the **vermis** of the cerebellum. Present in most vertebrates, the cerebellum functions principally for the coordination of motor activity. While the cerebrum initiates the movement, the cerebellum will coordinate all the components so the movement is smooth. However, new research is shedding light on other activities that the cerebellum may be involved in, such as learning, mood, and behavior.

### Brain Stem

The posterior portion of the brain, which is connected to the spinal cord, is termed the brain stem and comprises the **pons** and the **medulla oblongata**. The pons is the enlarged beginning of the brain stem, and is located where nerve tracts that enter and exit the cerebrum are organized. Additionally, the pons has regulatory centers for breathing and heart rate. Extending downward and composed of the nerve tracts to and from the spinal cord is the medulla oblongata. This region also serves as a critical regulator of basic body functions such as respiration and heart rate. The brain stem continues to function in individuals who have clinically been determined to be "brain dead."

## Spinal Cord

The spinal cord is the inferior extension of the CNS throughout the axial core of the body. Surrounded by the vertebral column, the spinal cord will receive and send nerve signals from and to the body, and transmit that information to and from the brain.

### Gray Matter

In the brain, the gray matter (nerve cell bodies) is located in the outer cortex, while the white matter is in the deeper portions. The white matter makes up the outer layers of the spinal cord, while the gray matter is in the center of the cord and is shaped much like the letter "H" or a butterfly. Divided in half, much like the cerebrum, the spinal cord has a gray matter connection that allows fibers to pass from one side to the other called the **commissure** (the middle part of the letter H in the example). In each half of the cord, the remainder of the gray matter is present as the dorsal horn (top of the long arm of the letter H) and the ventral horn (at the bottom). The dorsal horn is where you find synapses between sensory neurons (bringing in information from the periphery) and interconnecting neurons that will transmit the signal elsewhere. The bodies of the **interneurons**, neurons that only function to interconnect other neurons, are located here. Cell bodies of the motor neurons (sending information out to the periphery) are located in the ventral horn of the gray matter.

### White Matter

The white matter surrounding the gray matter of the spinal cord consists of the **myelinated axons** of nerves that carry sensory information to the brain (i.e., the ascending or afferent fibers) or motor information from the brain (i.e., the descending or efferent fibers), down the cord to go out to targets in the body. These nerves will form synapses with nerve cell bodies in the gray matter of the spinal cord before proceeding either up or out of the cord.

## Peripheral Nervous System

The nerves that are present outside of the CNS are collectively termed the **peripheral nervous system** (PNS). Most will emanate from the spinal cord and carry both sensory and motor information to and from the CNS. However, others will come from the brain or brain stem as **cranial nerves** to service the needs of the body.

### Cranial Nerves

Twelve cranial nerves are present in the human body and are numbered 1–12 (usually with Roman numerals). The first 9 nerves function with all of the special senses or control movement of the face and eyes. Cranial nerve X (10), the vagus nerve, serves functions and organs throughout the body and is the most widespread of the cranial nerves. Nerves XI (11) and XII (12) control head turning and tongue movement, respectively.

**What does *vagus* mean?**
*Vagus*, from Latin, means "wanderer" and perfectly describes the wide distribution of this nerve throughout the body.

## Spinal Nerves

Composed of sensory and motor fibers, spinal nerves travel laterally from the spinal cord and course throughout the body. There are 31 pairs of spinal nerves, each attached at the spinal cord by a dorsal root and a ventral root. The cell bodies in sensory neurons are located just peripheral to the spinal cord in collections of cell bodies called **ganglia**.

From the dorsal root ganglia, sensory cells send axons into the dorsal horn of the gray matter to form synapses with interneurons of the spinal cord. Motor nerve cells will extend their axons away from the spinal cord via the ventral root, which will eventually connect with the dorsal root and become the full final nerve. Think of the connection between the two-way spinal nerve and the spinal cord in much the same way as a street ends in a cul de sac. The street has lanes running in opposite directions much like the sensory and motor nerves have information moving in opposite directions. The turn-around of the cul de sac would be much like how the neurons and their signals are relayed in a curved manner, just like cars drive around the curve to go in the opposite direction.

# Autonomic Nervous System

The CNS is organized into separate control centers for the sensory information and motor output. The motor side is divided into the **somatic** portion (control for skeletal muscles) and the **autonomic** (subconscious control of smooth and cardiac muscle as well as glands). This division is further subdivided based on anatomy, neurotransmitters used, and physiological effects that result.

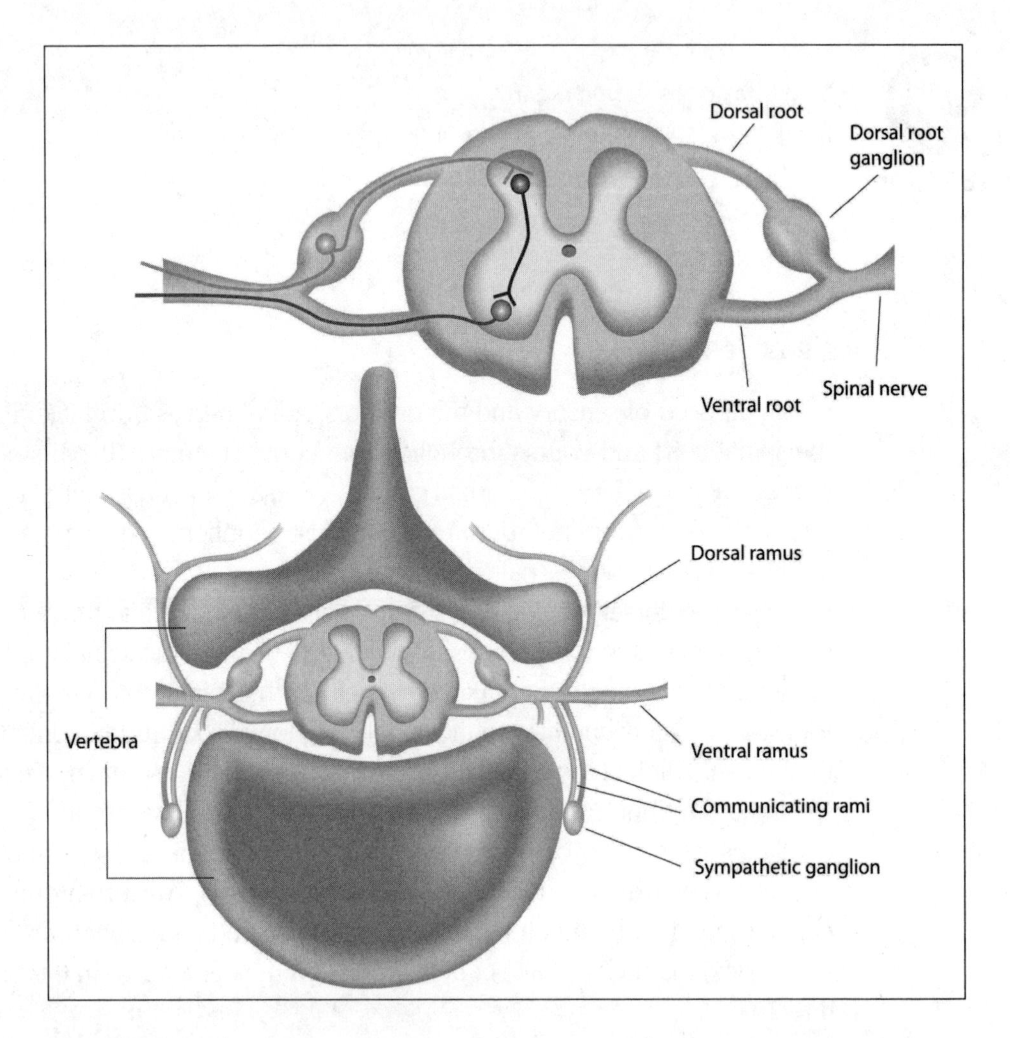


This illustration shows the arrangement of the sensory and motor neurons in a spinal nerve and their relationship to the spinal cord. Also notice the interconnection of these two neurons on the interior of the spinal cord that is mediated via the interneuron.

## Autonomic Reflex

Unlike the knee-jerk reflex, which consists of one sensory nerve and one motor nerve for an immediate and subconscious reaction, the autonomic reflex involves a series of 2 motor neurons that will deliver motor instructions to target tissues. The first motor neuron will leave the ventral horn of the spinal cord as all other nerves do; however, before reaching a target, it will form a synapse with a second neuron in the periphery of the body. These 2 nerve cells are named relative to the ganglion: the **preganglionic nerve** and the **postganglionic nerve**. The postganglionic nerve will innervate the target tissue and elicit the response.

Since the preganglionic nerve will always excite the postganglionic nerve, the neurotransmitter used by all preganglionic nerves (to signal the postganglionic nerve) will be acetylcholine, in both sympathetic and parasympathetic systems.

## Sympathetic System

Also known as the fight-or-flight nervous system, the sympathetic system will instantly prepare the body for intense physical activity. Most people have experienced this in those moments immediately following a frightening event, such as a near automobile accident, when your heart is racing out of control and your grip on the steering wheel is incredibly strong. The effects of this response are facilitated and initiated by neurotransmitters and a related hormone, **epinephrine** (adrenaline). Using the blood stream, this hormone can quickly spread throughout the body and cause the entire body to prepare for action.

Preganglionic sympathetic nerves can be located emanating from the spinal cord between the first thoracic vertebra (T1) to the second lumbar vertebra (L2). These fibers are said to form in the thoracolumbar region.

As soon as the preganglionic nerves leave the spinal cord and become part of spinal nerves, they will exit the spinal nerve via a pathway called a **white ramus**. This is not unlike a car exiting an interstate highway to gain access to a side road. In this case, the nerve exits the spinal nerve to gain access to one of the ganglia that formed an interconnected chain that mirrors the spinal cord. This sympathetic chain of ganglia allows the dispersal of nervous system information along a wide stretch of the thoracic and abdominal region, and controls activities such as shutting down much of the digestive system during a sympathetic response.

From the synapse in the sympathetic ganglia, the postganglionic nerves will enter the spinal nerve via a gray ramus (i.e., the on-ramp) and will run to and innervate the target tissue. However, some nerves will bypass the chain of ganglia and form 2 large nerve bundles, the greater and lesser **splanchnic** nerves. These will course through the body and form synapses with postganglionic nerves in collateral ganglia, such as the celiac, superior mesenteric, and inferior mesenteric ganglia, before extending to the target tissue.

## Neurotransmitters

Norepinephrine is the neurotransmitter used by sympathetic postganglionic nerves to stimulate their target tissue into action (or inaction). There are a few postganglionic neurons that use acetylcholine, such as the nerves going to sweat glands to stimulate perspiration.

## Effects

The physiological effects during a sympathetic response rev up the body in preparation for intense activity. These include increased heart and respiratory rates, dilation of the pupils, increase in blood pressure, and the direction of more of the body's blood supply to the skeletal muscles. The redirection of blood occurs partly from the dilation of blood vessels to the muscles; however, much of the shift is due to the constriction of blood to the digestive and urinary systems. Systems deemed nonessential for immediate survival are restricted in blood supply (to a minimal level) to divert those resources to the organs and tissue needed for short-term survival.

## Parasympathetic System

The opposite of the sympathetic system, the parasympathetic system is in control during those periods of rest where the body can perform nonactive functions such as digestion. Thus, this system is often referred to as the rest and digest system.

## Anatomy

The preganglionic fibers exit the spinal cord in the areas not used by the sympathetic system: the cervical and the sacral regions. Sympathetic nerves are in the middle, and the parasympathetic nerves are at the top and bottom of the spinal cord. In the head region, cranial nerves will serve as preganglionic neurons for the parasympathetic system. These will run practically the entire distance to the target before forming the synapse with postganglionic neurons in what are termed **terminal ganglia**. Other nerves will in fact terminate within the organ itself.

Terminal ganglia get their name because they occur so near the "end" of the path to the target tissue. They control functions such as tear production and saliva secretion.

The single neurotransmitter used in the parasympathetic nervous system is acetylcholine. Its effect on the body is the result of signal transduction from muscarinic ACh receptors and their downstream targets.

The physiological effects of the 2 autonomic systems are the opposite of each other. For example, heart rate increases during a sympathetic response and decreases in a parasympathetic response.

Since this system is the "rest and digest" controller, those activities will be dominant. Glandular secretions of the GI tract, peristaltic movement, and absorption in the lower alimentary canal will occur. Other physiological criteria will not be directed to fulfill specific roles as much as they will be turned down to conserve energy. Heart and respiratory rates will decrease as will the amount of blood going to skeletal muscles.

## Diseases and Disorders

Problems with the nervous system (either CNS or PNS) can lead to profound muscle dysfunction as well as cognitive, behavioral, and social disorders. While there are too many to explain in this unit, some of the more familiar disorders will be described.

## Parkinson's Disease

Parkinson's disease is a disease resulting from lack of dopamine production or from the failure to detect this essential CNS neurotransmitter. Typically manifesting in adults over 50 years of age, symptoms begin with motor dysfunction including tremors, stiffness, and problems with walking. These early symptoms will eventually lead to cognitive and social difficulties such as dementia as the disease progresses.

There is no cure for Parkinson's disease; much research has looked into disease prevention and management of the symptoms. Several studies have reported links between antioxidants (such as vitamin C) and the prevention of Parkinson's disease; however, this remains unresolved.

For patients who exhibit a decrease in dopamine production, medications that allow the neurotransmitter a longer active life are effective at reducing the symptoms. Typically, when dopamine is secreted toward synapses, an enzyme called **monoamine oxidase** (MAO) will break down dopamine to stop the signal (in order to allow another to occur if signaled). MAO inhibitors will reduce the efficacy of the enzyme, prolong the active life of dopamine, and mimic the presence of greater amounts of dopamine.

## Alzheimer's Disease

Accounting for approximately 60–70 percent of dementia cases, Alzheimer's disease is a neurodegenerative disorder primarily affecting adults over the age of 65. Although the cause is poorly understood, heredity is thought to play a large role in determining who contracts the disease.

Several hypotheses have tried to explain the cellular/molecular events that lead to the disease; however, the leading hypothesis focuses on the role of amyloid plaque proteins (APP) and their extraneous accumulation in the area surrounding neurons. It remains unclear if APP causes Alzheimer's, since vaccines that remove the APP accumulations (plaques) do not reverse the dementia in these patients.

Initial symptoms of Alzheimer's include short-term memory loss that, in time, will lead to more severe cognitive disorders. Disorientation and problems with language that may cause the individual to isolate himself are not uncommon. Further advancement of the disease will lead to loss of systemic bodily functions and death.

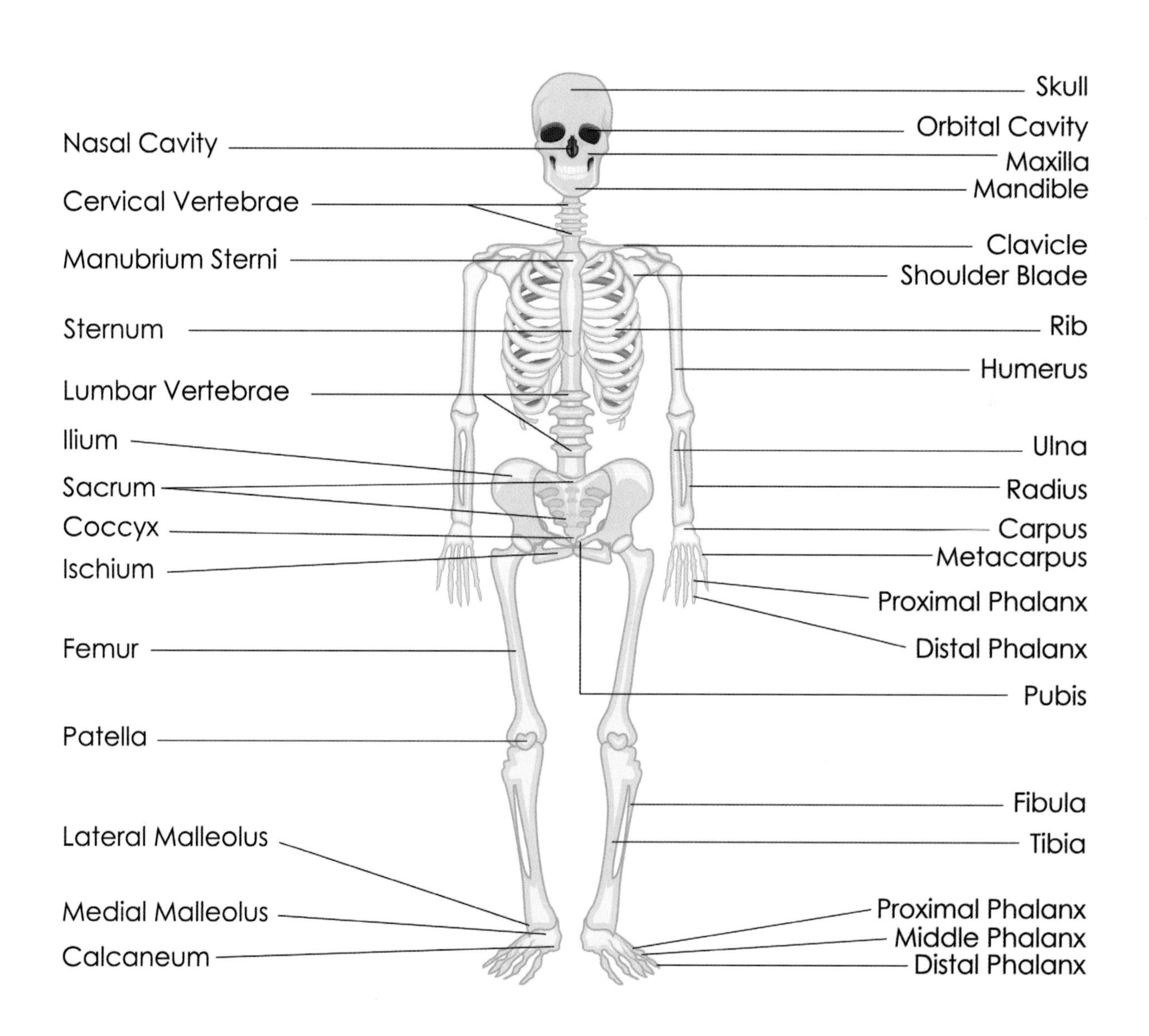


## Skeletal System

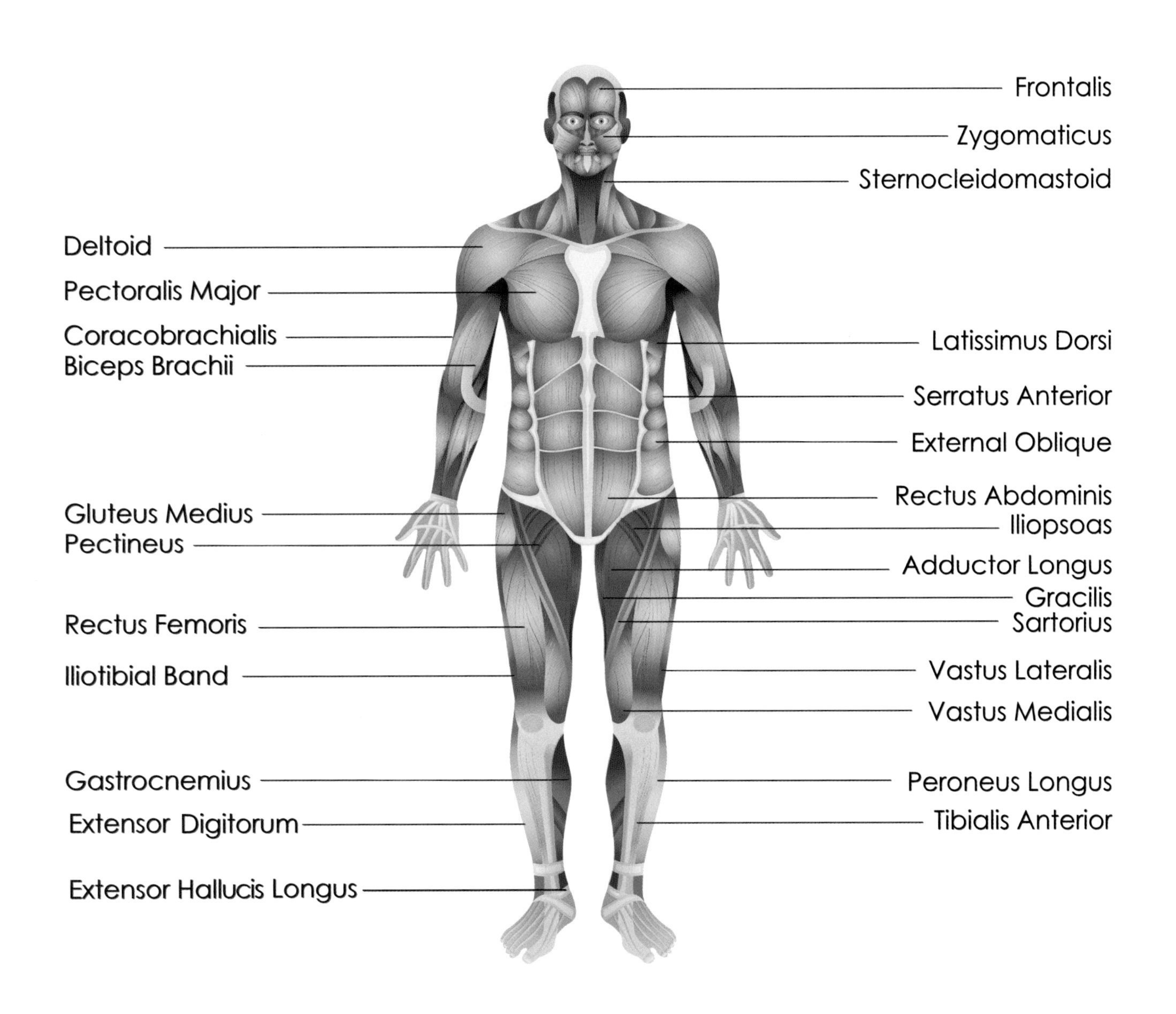


## Muscular System

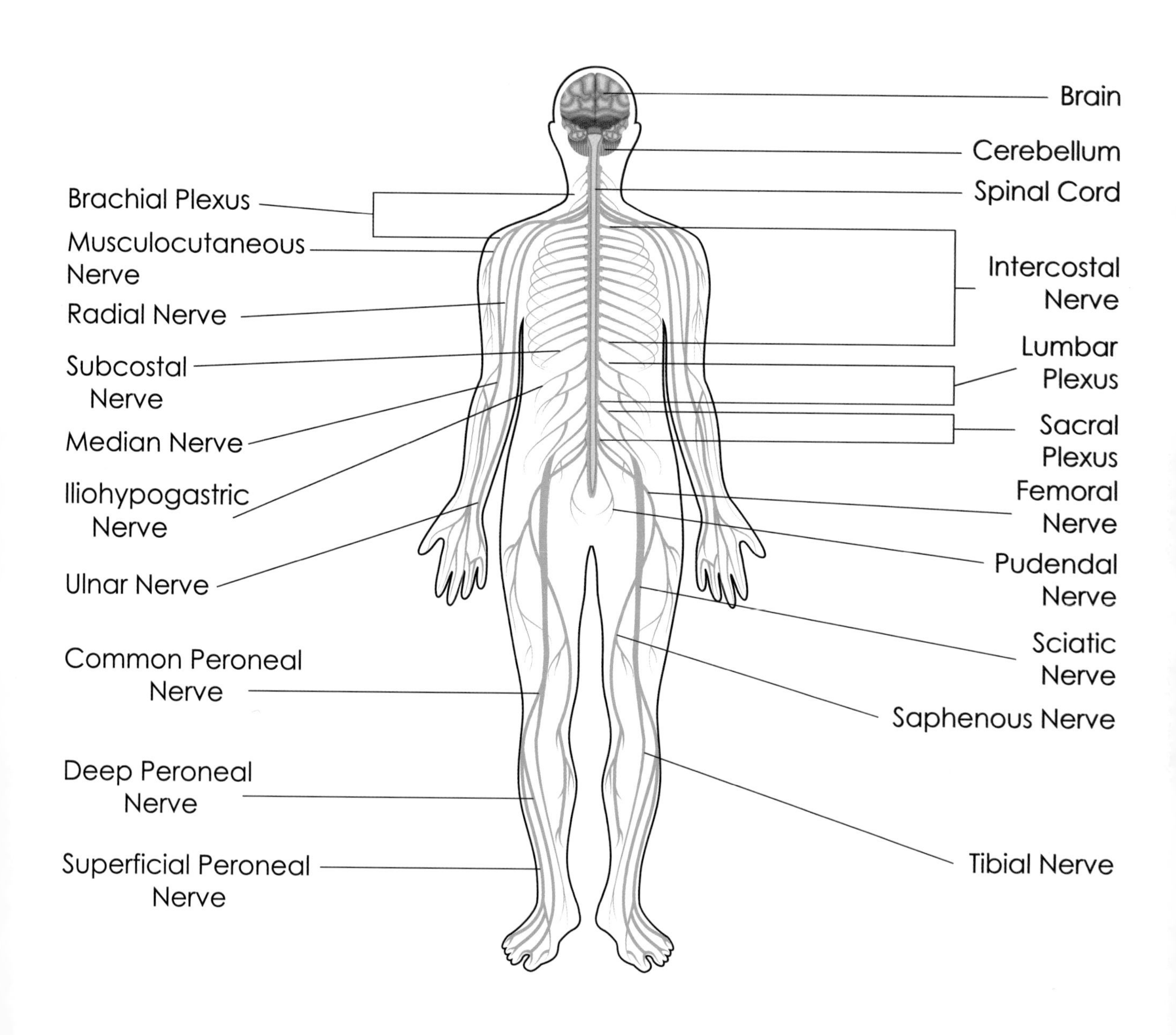


**Nervous System**

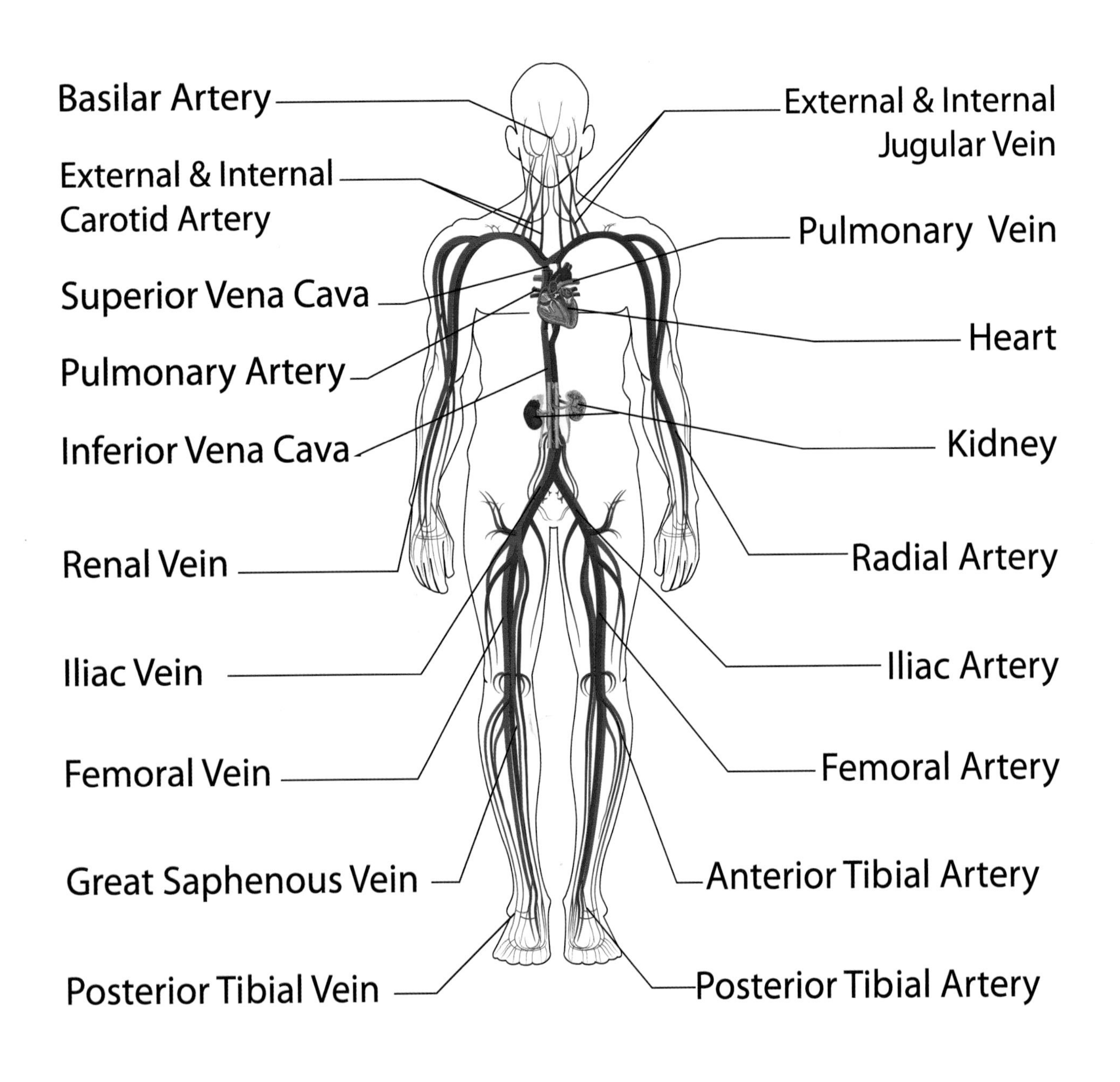


Circulatory System

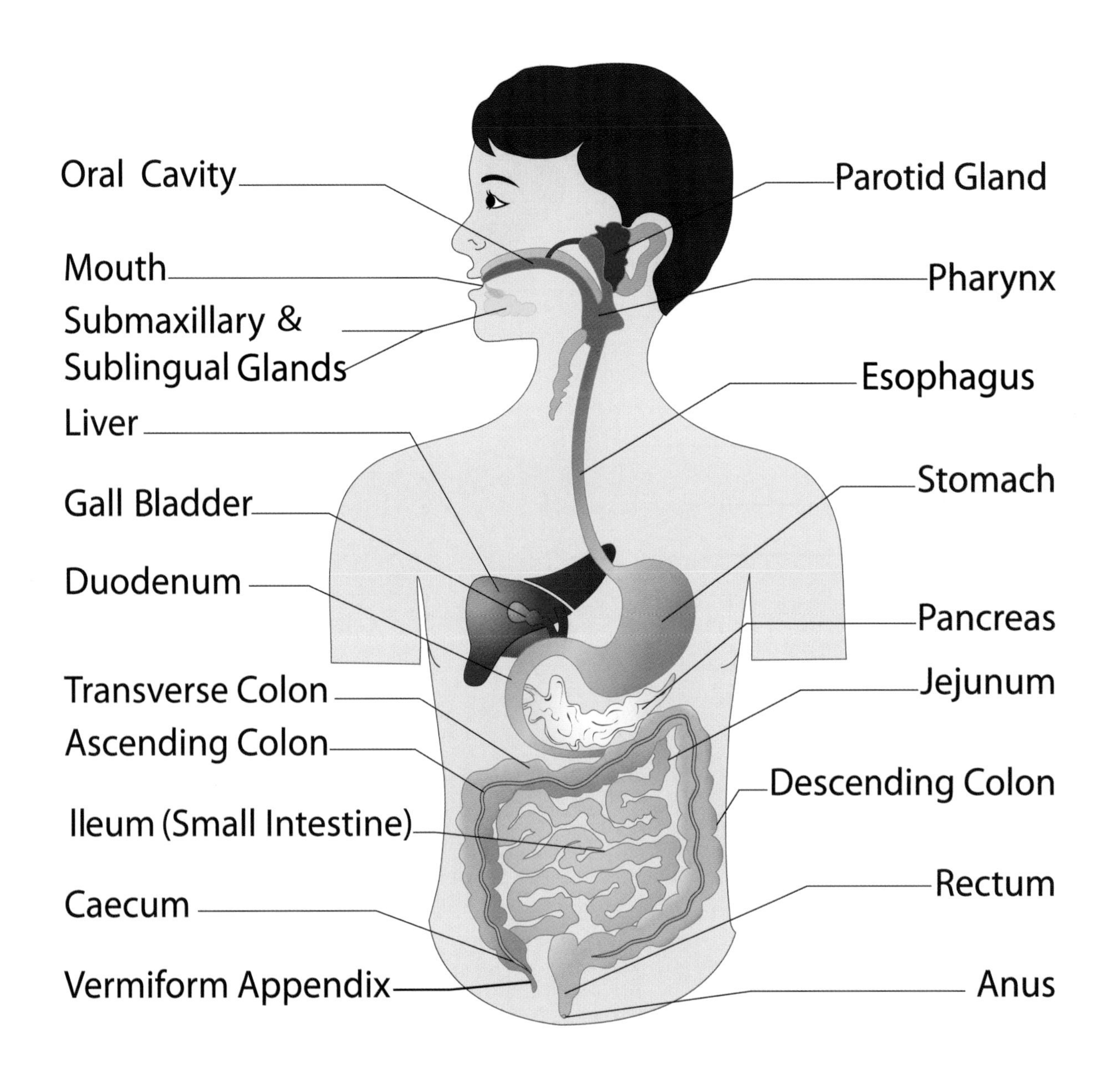


Digestive System

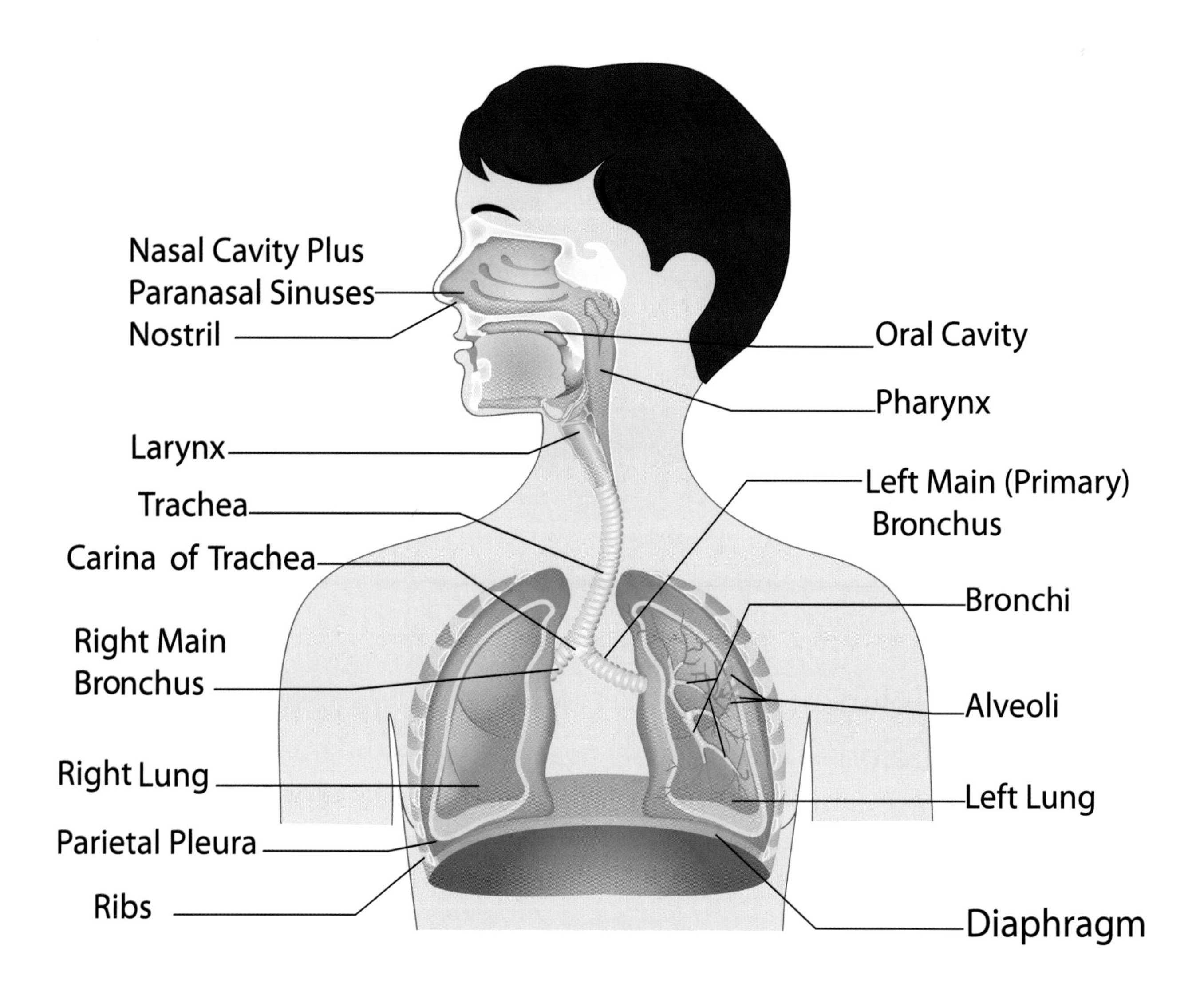


Respiratory System

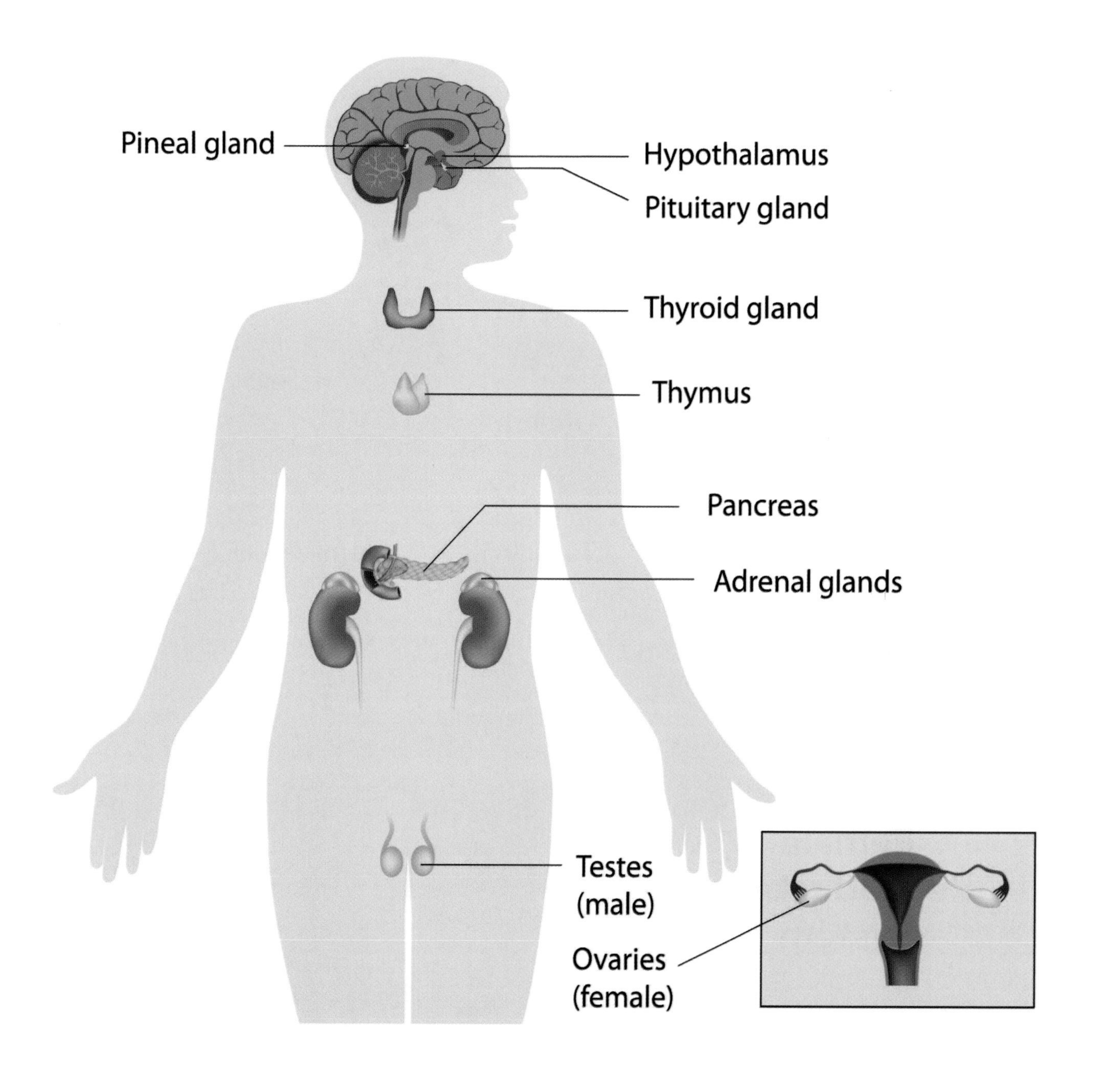


**Endocrine System**

## Male Reproductive System

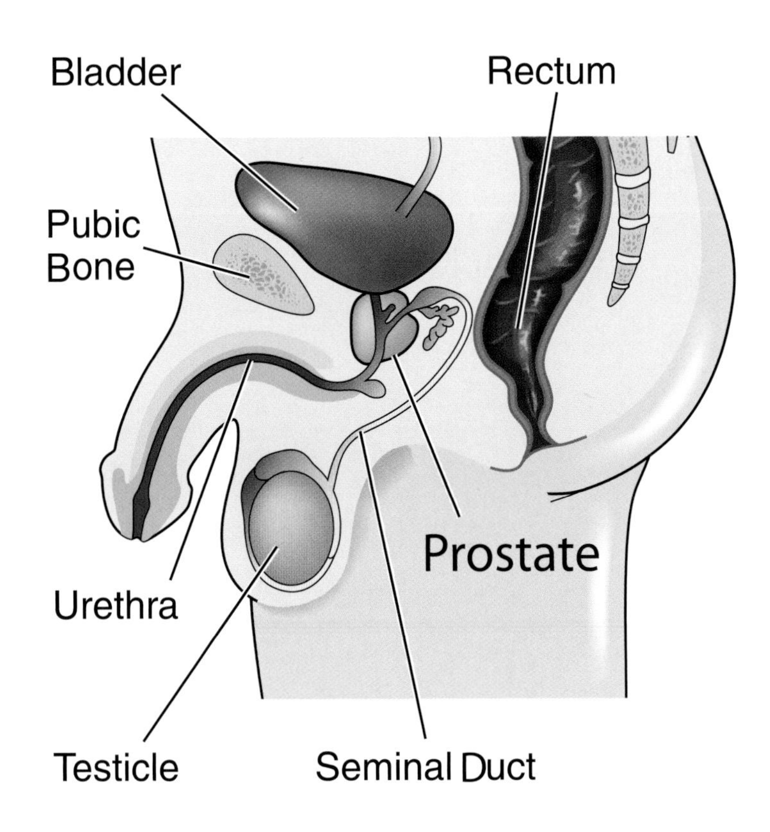


## Female Reproductive System

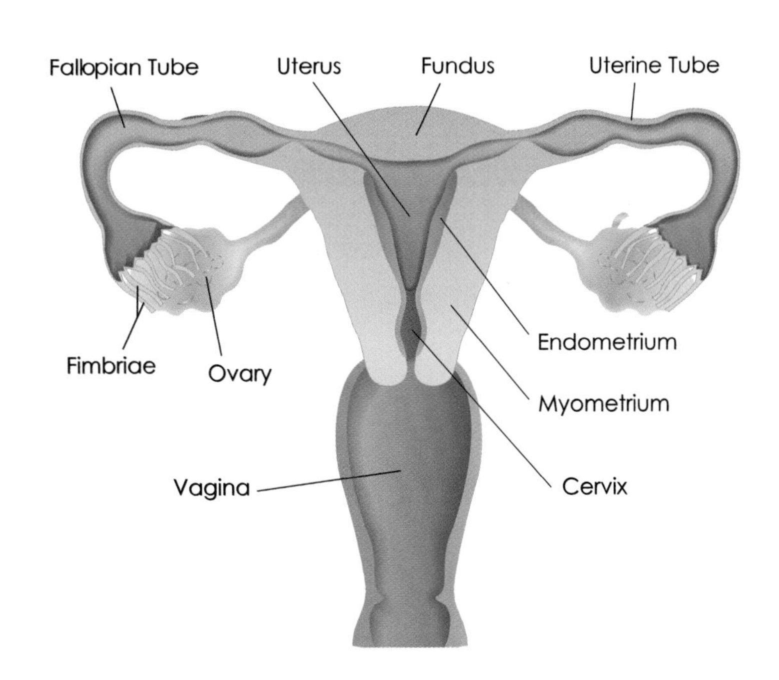

## Attention Deficit Hyperactivity Disorder

A neurodevelopmental disorder in children, attention deficit hyperactivity disorder (ADHD) or attention deficit disorder (ADD) affects a child's ability to focus attention for more than a few minutes on tasks, with or without increased extraneous motor activity. These children are often identified once they begin socialized education (either preschool or primary school) because of their failure to stay on task and/or inability to sit quietly for any length of time. While some children diagnosed with ADHD or ADD simply may not have been socialized properly or taught acceptable behaviors at home, others do have a pathophysiological basis for these disorders and can benefit from counseling and medication.

Stimulants are the most popular and effective means to treat children with ADHD or ADD. The increased level of brain activity these stimulants cause may actually help streamline information and result in greater focus and attention, rather than allow the slower information to lead to distraction. However, this may not always be the best solution. These individuals must learn to control their behavior as they move into early adulthood, and most will be able to withdraw from the medication.

## Autism

This disease begins to manifest itself in early childhood and can be seen in a delay in language development as well as social dysfunction. However, the most significant symptom is repetitive behaviors, which can be simple task repetition or may be a compulsive action or even ritualistic in nature. These behaviors help in the distinction between autism and ADHD or ADD.

While many neuronal disorders have pathophysiological or chemical foundations, the underlying malformation of autism isn't well understood. One possibility is the malformation of synapses throughout the CNS. Such a malformation could delay or incorrectly relay signals throughout the nervous system and possibly lead to specific synaptic pathways that are comfortable for the individual. These patterns are repeated frequently, and it becomes difficult to break the repetitive cycle, much like an individual becoming addicted to a drug that stimulates the pleasure center of the brain.

One area related to the causes of autism that has been highlighted in many media articles is the relationship between childhood vaccinations and autism. To date, the link between the two has not been widely accepted within the medical and basic scientific community.

CHAPTER 9

# Sensory System

Just as astronauts in space monitor the conditions both inside and outside their ship, as well as where their ship is located in space, the human body has an intricate system of sensors that provide information back to the brain to elicit a response, which may be either conscious or subconscious, to provide better conditions for health and survival. Whether it is the reaction to touching a hot object or a matter of feeling thirsty, the sensory system is working 24/7 to keep your body functioning properly.

## Reception and Perception

As you look at the different means by which your body obtains information about your interior and exterior environment, it becomes increasingly clear that this is a 2-step process. The information must be obtained via specialized cells that can detect stimuli and it must be interpreted in context by the brain before a response can be made.

Receptors in your body provide a vast array of information to the brain, such as internal and external temperature, blood pressure, light, sound, taste, smell, as well as balance and body position. If any of these receptors fails to function, the information will not reach the brain and will not be detected. Therefore, no response, either reflex or conscious, will be made. However, the brain may have established patterns of dealing with such sensory information that continue even after loss of input. For example, amputees often relate experiences of "phantom limbs" to their physicians. Although the limb is gone, an amputee's brain will still provide signals that make her feel as if there is touch, pain, or temperature being sensed on that limb. This shows the power of the brain and its importance in the perception of a stimulus.

In a similar way, information from receptors must be understood appropriately by the brain to be perceived. Tastes, smells, and images of color must be learned in the early years of life and assembled into a library of sensory information that can be sorted through when encountered again. If the brain never develops these early patterns because it was deprived of sensory input until later in life, the brain may never be able to accurately interpret the information and the individual will be unable to perceive certain stimuli. For example, individuals who have been blind from birth because of eye problems that are repaired later in life often have a difficult time with depth perception.

It takes both aspects of the sensory system working in unison for an individual to adequately function and monitor the internal and external world in which we exist and survive.

## Special Senses

To understand the organization of the sensory system, the sensory receptors are categorized into groups based on how localized or widespread they occur in the body. Therefore, those that are localized in a single location are referred to as **organs of special senses**. This group includes the cells and tissues required for smell, taste, vision, hearing, and balance. In the past, touch was included rather than balance in the "5 senses." However, much of the external surface of the body can detect touch as pressure. For that reason, touch was moved into the category of general senses that will be covered later in this unit.

## Smell

The first of the special senses and one of the two senses that detect chemical signals, the sense of smell, or **olfaction**, begins in the lining of the nasal cavity. On the surface of the nasal epithelium, secretory cells produce and coat this area with a watery, protein-rich fluid that traps chemicals that are detected as odors as they are brought into the nasal cavity during inhalation.

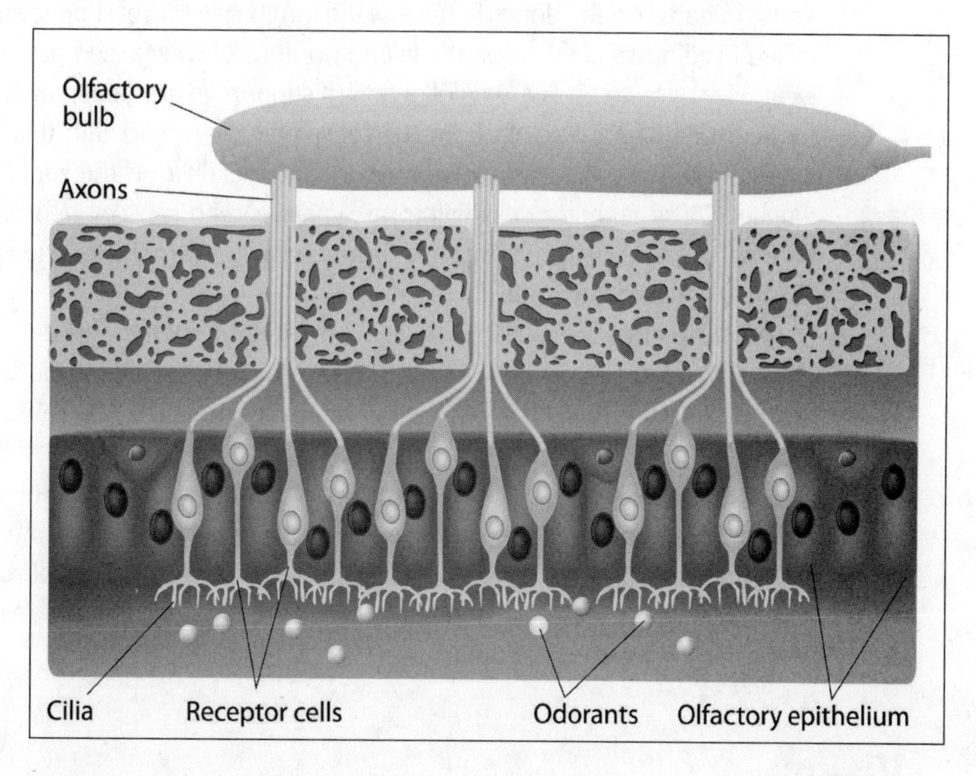


Olfactory receptors in the nose are responsible for detecting and relaying information to the brain that is perceived as smell.

Once they are trapped in the watery surface zone, these chemicals are detected by the olfactory receptor cells. These cells are an important member of the epithelial layer, and more readily resemble neurons than epithelium. On their surface and projecting into the watery chemical-filled fluid are **dendrites** that are often called cilia. These will increase the surface area where proteins can bind to the chemicals and begin a signal transduction pathway that leads to an electrical signal that will be sent to the brain. At the base of these cells, axons transmit these electrical signals on to mitral cells located in what is called the **olfactory bulb**. In the region of the forebrain, this is then connected to the olfactory centers of the brain via the olfactory tract.

## Taste

Taste (**gustation**) is the second of the chemically mediated specialized senses and is initiated on the dorsal surface of the tongue via barrel-shaped structures called taste buds. Located on the lateral portions of specialized papillae (projections), the taste buds are a collection of supportive and receptor cells used for taste. Hairlike microvilli from the receptor cells project into this opening to collect the chemicals dissolved in the watery saliva on the surface of the tongue. These receptor cells will form synapses with sensory neurons at the base of the taste bud and transmit information away from the tongue and on to the taste center of the brain.

All taste buds are capable of detecting several chemical stimuli, but are thought to specialize in one or two. They are distributed across the surface of the tongue and are only restricted by the locations of the papillae on which they are located. There are now 5 basic tastes that are mediated by receptors and require distinct signal transduction mechanisms to convey the stimulus and transduce it into electrical energy. Sweet, salty, sour, and bitter are 4 of the basic tastes that have been identified for decades. While most people are familiar with these, the newest of the basic tastes is umami, which is a savory flavor and common in Asian cultures.

## Vision

This highly complex system is designed to detect light energy and transduce it into electrical information that is sent to the visual cortex of the brain for perception.

Interestingly, light input in the right eye is perceived in the left part of the brain and vice versa. This crossing of the information and the slightly offset position of the eyes allow the brain to perceive different dimensions, including depth perception.

## Anatomy of the Eye

The eyes are organs that capture and focus light energy on the back portion of the eyeball where the **retina** is. The retina contains the photoreceptor cells. The **sclera** forms the outer covering of the eye and can be found in the front as the **cornea** (clear portion of the eye), which allows photons of light to pass into the eye. Under the cornea is a spherical diaphragm that can open or close to regulate the amount of light that will pass into the next compartment. This colored portion of the eye is called the **iris**, and its opening, which is black because it is dark in the back of the eye, is the **pupil**.

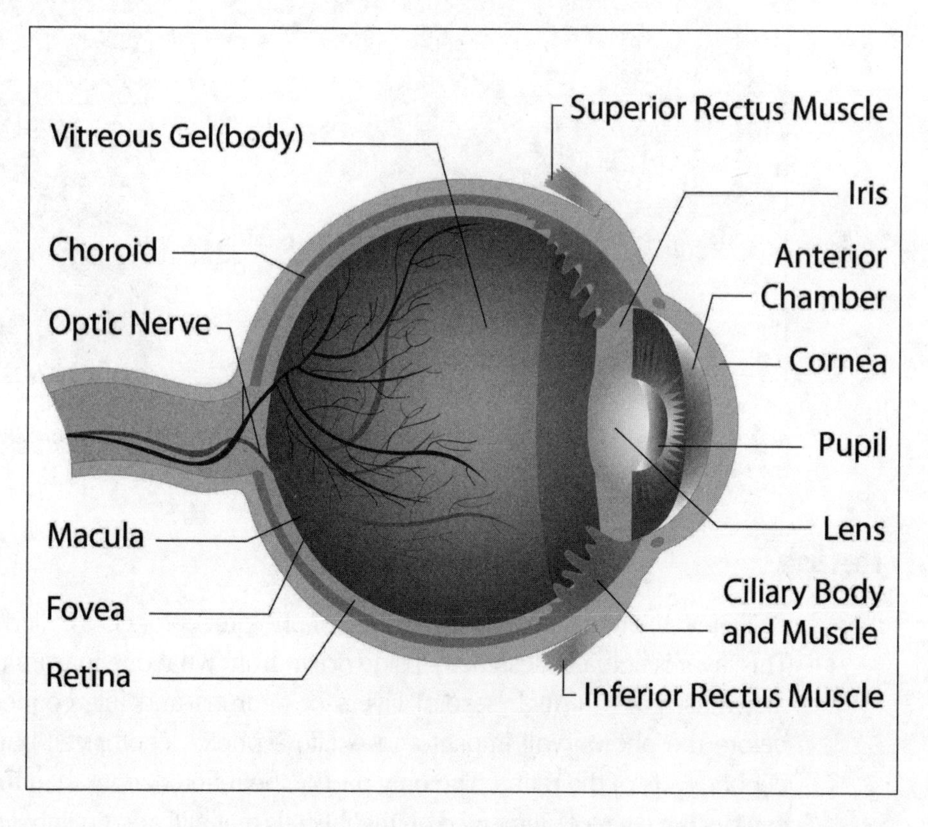


Light enters the front of the eye and passes through the cornea, through the opening of the iris called the pupil, and through the lens. Light then travels to the gelatinous material in the vitreous body before being detected by the retina in the back of the eyeball.

Suspended behind the iris and positioned immediately behind the pupil, the lens is kept in place by the ciliary muscle, which can apply tension to the lens to stretch it or allow it to contract in order to change the focal length of the light entering the eye (i.e., permit focal accommodation to see objects up close and then far away as the focus of the eye automatically adjusts). After light enters the eye and is focused by the lens, it passes through the large chamber of the eyeball that is filled with a gelatinous material called **vitreous humor** before passing into the photoreceptor layer: the retina.

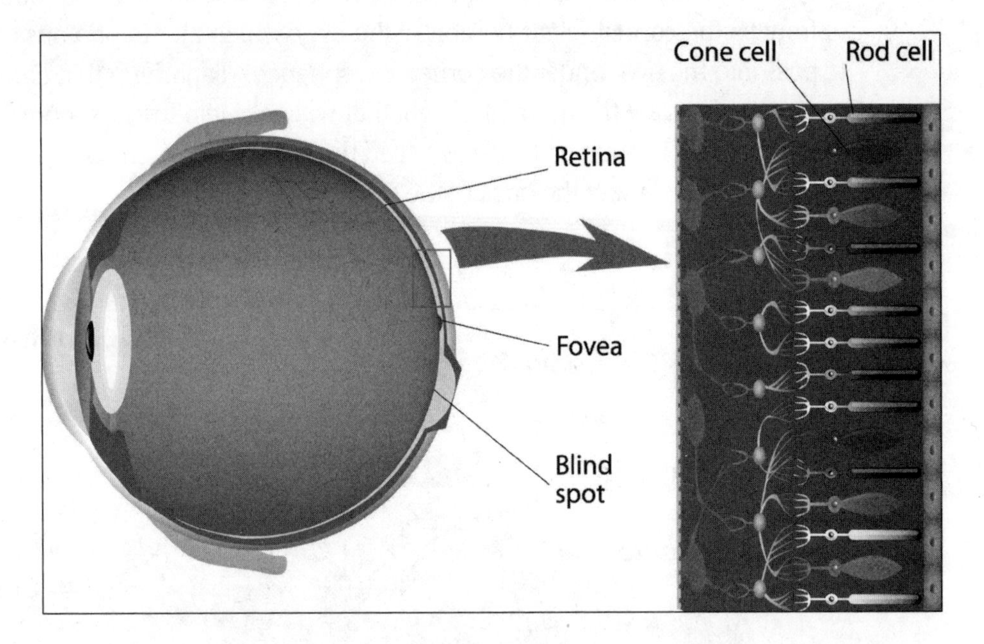


A cross section of the human eye, with a closer look at the photoreceptor cells present in the retina.

## Retina

The back of the eyeball has a layer of photoreceptor cells called the retina. This layer is actually designed upside down from what one may imagine. Light first must pass through several layers of neurons and interconnecting cells before the photon will impact and excite a photoreceptor cell buried in the deeper parts of the retina. The only part of the retina deeper than the photoreceptors is a layer of pigmented epithelial cells that will absorb light and prevent its reflection to reduce extraneous light and improve fine detail.

Nocturnal animals have a reflective pigment in their epithelium (the tapetum lucidum) that reflects light directly back onto the photoreceptor, thus stimulating it multiple times with a single photon and improving vision in low light.

## Photoreceptors

These cells can be found in the deep retina and contain a photo pigment that, when activated by a photon of light, will change shape and lead to a signal transduction cascade, ultimately generating an electrical signal in a neuron that will signal the visual cortex of the brain.

One type of photoreceptor is the **rod cells**. Shaped much like a comb for your hair, these rows of folded membranes that resemble the teeth of the comb contain the photo pigment rhodopsin, which, when activated, will lead to the change in the shape of the larger molecule where it is attached (opsin). An essential component of this photo pigment complex is a molecule called **retinol**, which is related to vitamin A. Rods are the more sensitive of the 2 types of photoreceptors and are responsible for vision in low light, which many have referred to as black and white vision. Also concentrated at the edge of the retina you will find the cones. Cones are essential for peripheral vision.

**Cone cells** are the second type of photoreceptor and are responsible for color vision in the retina. These cells also have folds of photo pigment, but are shorter than the rods and have a tapering or "cone" shape, hence the name. Three different cones are based on their photopsin (pigment) and the wavelength of light they detect. They are the red, green, and blue cones. Which cones are stimulated and to what degree determines the color of light that is signaled and perceived. This is not unlike your television screen or computer monitor where the same 3 colors are used to blend into any visual color in the spectrum. While cones are much less abundant than rods in the retina in general, in the **fovea** of the retina (the central part of the retina responsible for sharp central vision), cones greatly outnumber rods.

# Hearing

The last 2 special senses utilize mechanical energy. Sound waves (i.e., mechanical energy) move through the air and are funneled into the auditory tube via the external ear (**pinna**). This canal and the eardrum (**tympanic membrane**) at the deep end make up the outer ear. As waves impact the eardrum, the membrane vibrates, and energy is transferred through a series of small bones in the middle ear, or the chamber on the opposite side of the eardrum.

The **incus**, **malleus**, and **stapes** (anvil, hammer, and stirrup, respectively) transfer the mechanical energy of the sound waves from the eardrum to the **cochlea**, the organ of the inner ear that houses the auditory receptors. The stapes is attached to an oval window on the body of the cochlea. Waves are transferred into the fluid-filled tunnel of the coiled cochlea.

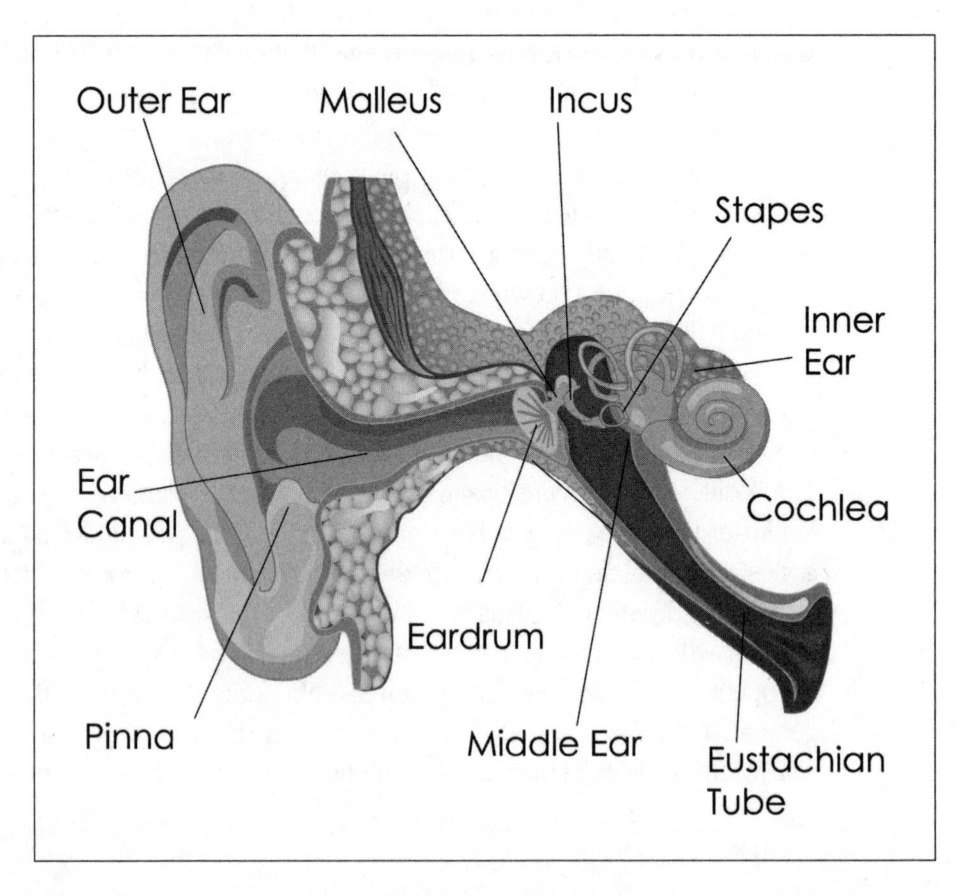


Illustration of the auditory system showing the components of the outer, middle, and inner ear.

The cochlea resembles a snail shell. The long fluid-filled tube, in which the sound waves will pass, is packed inside. This tube will be divided into an upper and lower chamber by the sensory **organ of Corti**. The very end of the tube will remain open and allow the sound waves in the upper chamber to run the full length of the tube. In doing so, the waves coursing through the fluid will compress a membrane in the organ of Corti, which will then compress down upon receptor cilia to cause an electrical signal. Since these receptors (and the organ of Corti) run the full length of the tube, the region where the receptors are stimulated will correspond to the wavelength and distinguish low from high tones before signaling the auditory center of the brain. This coiled nature of the organ of Corti can be compared to the coiled tubes of a French horn. Air moves around the tubes, and, depending on which valves are pressed, directing the air to a shorter or longer path, the tone is affected.

## Balance

Balance, which was only added to the list of special senses in the last several years, also has receptors located in the inner ear and is associated with the cochlea. While sound is detected in the body of the cochlea, balance is monitored in 3 semicircular canals that are attached to the body of the cochlea. Oriented in 3 planes, which will detect movement in 3 dimensions, these canals respond to acceleration and deceleration and utilize the law of inertia to function.

The law of inertia states that an object at rest will stay at rest and an object in motion will stay in motion until acted upon by an opposing force.

Fluid called **endolymph** fills each semicircular canal and, much like any object, will move relative to the body and follow the law of inertia. Receptor cells project cilia upward and into this fluid. At rest, the cilia are standing straight and no signal is being sent to the body. However, when the body starts to move, the receptor cells (and cilia) will accelerate at the same rate as the body. But the endolymph will remain stationary for a brief moment, just long enough to bend the cilia of the receptors and transduce an electrical signal that motion has been initiated. Likewise, when the body stops, the fluid will remain in motion and bend the cilia in the other direction and the brain will perceive this as halting. With these 3 canals arranged in 3 planes, motion in any of the 3 dimensions will be detected.

## General Senses

Unlike the special senses, receptors for the generalized senses are spread throughout the body. The detection of temperature, pressure, pain, and body position occurs over much of the body surface and provides an abundance of sensory information to the brain.

### Touch

You can find the receptors for touch between the upper layer of skin (the epidermis) and the deeper layer (the dermis). Some receptors are adept at detecting changes in pressure (i.e., fast and slow receptors—fast may be more like painful pressure, whereas slow may be more like delicate touch), while some respond to vibrations or sustained pressure. Two of the more common receptors, **Meissner's** and **Pacinian corpuscles**, respond to slow and fast vibrations, respectively. Additionally, the Pacinian corpuscle will also detect deep pressure. These cells are specialized and encapsulated in such a way that the mechanical pressure will lead to a signal transduction cascade and neuronal signaling. This is how the brain perceives touch.

### Pain

Pain receptors (**nociceptors**) are free nerve endings that respond to various stimuli, which leads to the perception of pain. Extreme temperatures, excessive mechanical forces, and chemical damage are a few of the more common stimuli that are detected by these free endings as pain.

### Proprioception

**Proprioception** is the perception of the body's position in space. At all times the brain will know where the right arm is relative to the left arm. All body parts will be linked in such a way. This enables an individual to close his eyes and touch his nose with the tips of his fingers. This is possible by using the sense of proprioception. However, in cases of alcohol intoxication, in which the proprioception centers of the brain are impaired, this test becomes part of a sobriety examination. Encapsulated specialized receptors in the muscles (muscle spindles) and in the tendons (i.e., Golgi tendon organs) are a few of the receptors that detect body position based on muscle and tendon tension and relay information back to the brain. The brain then assimilates the information into a global image of where all the parts of the body are located at any specific instance.

## Diseases and Disorders

The ability for organisms to sense the world around them is essential for their survival and reproduction. For humans, this also affects the quality of life in daily activity. While defects in the ability to smell may not seem as debilitating as being blind, there are disadvantages to losing or having less effective mechanisms for any sense.

### Colorblindness

This condition results from an inadequacy, malfunction, or malformation of one or more types of cones in the retina. The most common type of colorblindness, red-green colorblindness, is when red and green color discrimination is decreased. These individuals, for instance, must be aware of the arrangement of the traffic lights since the red and green lights will appear to be the same color. Affecting 1 percent of males, this is a hereditary condition and is termed X-linked, meaning the gene that is the problem is on the X chromosome. Also, this is a recessive mutation in that the presence of a dominant gene can overcome the condition and the individual will only be a carrier of the colorblind gene. For example, a female has 2 X chromosomes. For a female to be colorblind, she must have both chromosomes containing the mutation. However, for a son to be colorblind, his mother must at least be a carrier of the mutation (he will get the Y chromosome from the father and an X chromosome from the mother). Additionally, if the father of a female is colorblind (meaning his one and only X chromosome has the colorblind gene), then she will at least be a carrier, and the X chromosome she received from her mother will determine whether she will be colorblind.

## Anosmia

**Anosmia** means a lack of the sense of smell. Smell and taste are intimately linked together to expand the repertoire of each sense. A person lacking the sense of smell has a much less discriminating palate compared to individuals with a functional olfactory system. Often, temporary anosmia may occur during a sinus infection, and that is usually why the taste of food is dulled during such a sickness.

## Vertigo

Vertigo is a result of complications with the functioning of the inner ear, specifically the semicircular canals. Often, this may be caused by ear infections leading to increased pressure and impaired function of the canals, or it can be caused with age, as the canals take on a less-than-optimal shape. The lack of information, or incorrect information, pertaining to balance can result in extreme dizziness, nausea, and vomiting, and increases the likelihood of falling. This is particularly problematic in older individuals.

## Motion Sickness

With some of the same symptoms as vertigo, motion sickness (car sickness, air sickness) results from crossed signals between what the visual centers of the brain interpret and what the motion centers of the brain perceive. There is confusion between what the brain thinks is happening versus what the balance centers of the brain think. For example, many people get seasick on a ship when they are inside of a portion of the ship that has no windows. Their visual cortex sees the floor, walls, and ceiling, and relative to the body there is no movement. However, because the boat is moving up and down with the waves, the body is in fact moving and the movement is being detected by the semicircular canals of the inner ear. The brain interprets this counterintuitive information as a toxin-induced hallucination and the response is to cause vomiting to try and eliminate the toxin from the body. The resulting symptoms of nausea, vomiting, and perspiring are a result of this defensive action. Closing the eyes and focusing on the motion often can be enough to alleviate the symptoms.

## CHAPTER 10

# Cardiovascular System

The cardiovascular system transports materials throughout the body that cells and tissue could not survive without. Conveying oxygen to the tissues and relaying carbon dioxide back to the lungs is an essential function of this system. However, other critical materials are also conveyed in the blood. Hormones produced by the endocrine glands utilize the blood stream as a means to travel to different parts of the body where their effects on growth, metabolism, and reproduction take place. Immune system cells also take advantage of this flowing pathway to rapidly gain access to areas of the body that are under attack from pathogens. The cardiovascular system is much more than merely a pump and plumbing to transport gases.

# Heart

The driving force of the cardiovascular system begins at the heart. This muscular organ begins pumping even before the heart is fully formed in the embryo and will consistently and spontaneously beat for a lifetime. Divided into 4 chambers and separated into right and left portions, the heart is a unidirectional pump that moves blood throughout the body while at the same time pumping blood to and from the lungs. The heart functions as a pump for 2 circulations: the **systemic circulation** for the body and the **pulmonary circulation** for the lungs. Using both of these will remove carbon dioxide from the body while supplying fresh oxygen to the tissues.

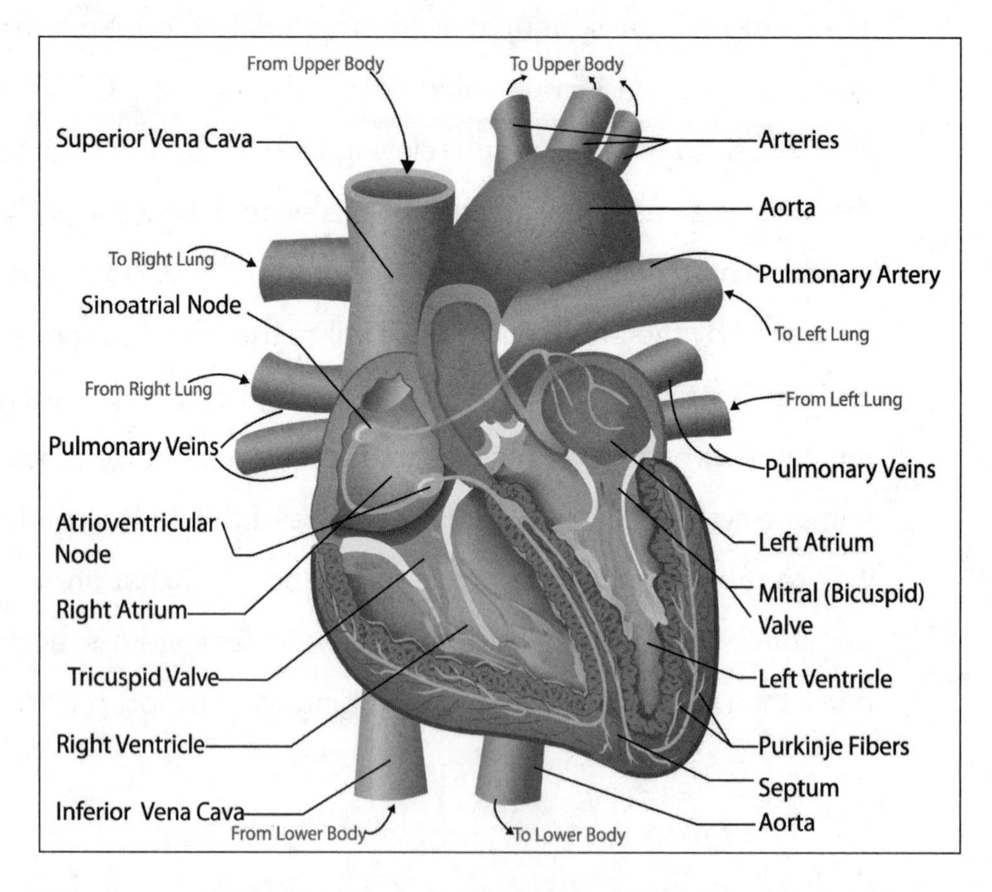


Valves connect the four chambers and regulate the flow of blood into the aorta and pulmonary trunk. Major vessels supply blood to the heart and convey blood away from the heart.

## Chambers

The 2 superior chambers of the heart are the **atria**, which actually begin as a single chamber during embryonic development and later form a partition (**interatrial septum**) to divide into right and left chambers. These thin-walled chambers are the primary holding areas for blood returning from the body (**right atrium**) and from the lungs (**left atrium**). Movement of the blood from the atria downward into the respective **ventricles** (the chambers that are essentially the pumps for the blood) is accomplished primarily from low pressure pulling the blood downward. This explains why so little muscle is present in the atrial walls. However, embedded in the wall of the right atrium is a critical tissue called the **sinoatrial node** (pacemaker), which will be discussed later.

Much more cardiac muscle is present in the walls of the ventricles. Like the atria, the ventricle begins as a single chamber that will be separated into right and left chambers by a thick muscular partition (interventricular septum). Oxygen-depleted blood in the right atrium is pulled into the right ventricle as the ventricular walls expand during the relaxation phase (the **diastole**) of a cardiac cycle. Likewise, oxygen-rich blood in the left atrium is pulled into the left ventricle. The left ventricle has the thickest of the muscular walls because the blood in this side must be pumped throughout the body and back to the heart, while blood in the right ventricle must only have sufficient pressure to make it to the lungs and back to the heart.

## Valves

The valves regulate the passage of blood from one chamber to the next as well as insure that the blood only flows in one direction. One set of valves regulates the movement of blood between the atria and ventricles (**atrioventricular valves**), while another set (**semilunar valves**) will prevent blood from flowing back into the heart from major vessels.

### Atrioventricular Valves

The structure of these valves is similar in both the right and left atrioventricular canals. Formed from thin connective tissue sheets (**cusps**), the **tricuspid valve** (on the right) and the **bicuspid valve** (on the left) will be pushed against the wall of the ventricles as the blood is pulled into the ventricles from the atria. When the ventricles contract (an action called **systole**), blood is compressed and forced upward in the ventricles. The most open path out of the ventricle is back up into the atrium.

However, as the blood moves up, it will be forced under the cusps, which then inflate, and moved upward toward the atrium. If the cusps were the only structural portion of the valves, they would simply be pushed aside into the atrium as the blood rushes upward. This does not happen in the human heart because attached to each cusp are cords (**chordae tendineae**) made from tough connective tissue. These cords are anchored into large bundles of cardiac muscle (papillary muscle) present in the ventricle that contract when the ventricle as a whole contracts. Thus, as the blood is moved upward because of the ventricular contraction, the papillary muscles are pulling downward on the cords with the same force as the upward-moving blood. Because these forces are equal, they allow the cusps to move upward only until they are aligned at the atrioventricular canal. Their meeting closes the canal off, preventing the blood from moving backward (regurgitating). The closing of these valves produces the first heart sound, or the "lub" that is heard for the heartbeat.

This structure is present on both the right and left sides of the heart. However, on the right side, the valve consists of 3 cusps and is aptly named the tricuspid valve. Consisting of only 2 cusps on the left, it is named the bicuspid or the **mitral valve**.

Atrioventricular valves can be compared to a skydiver. The cusps of the valves are the canopy of the parachute, the shock cords are the chordae tendineae, and the papillary muscles are the skydiver. In the heart, as blood inflates the cusps (and are being pushed backward into the atrium), the papillary muscles (much like the person) pulls down on the chords to prevent the valves from moving too far upward.

### Semilunar Valves

These valves, which prevent blood from falling backward into the heart when the ventricles relax, are actually present in the great vessels just outside of the heart. Blood from the right ventricle is forced into the **pulmonary trunk**, major vessel conveying blood toward the lungs, and is destined for the lungs.

Blood from the left ventricle is moved into the **aortic trunk**, major vessel conveying blood to the body, and is pushed to the rest of the body. Located within each of these vessels are 3 cusps that resemble pockets. In much the same way as someone puts money into his pocket, the only way to inflate these cusps is when the blood falls downward in an attempt to return to the ventricles. With 3 cusps positioned in the same plane, when all 3 inflate, it effectively closes off the passageway and keeps the blood in the vessel. Therefore, these valves are inverted compared to the tricuspid or bicuspid valves that allow blood to move into the ventricles and not out. The semilunar valves allow blood to exit the ventricles but not return.

# Heart Regulation

A unique feature of the heart is that it will beat spontaneously and rhythmically on its own. Regardless of any nervous system signals, the heart starts beating before you are born and will consistently and continually beat for your entire lifetime.

## Pacemaker and Conduction

Setting the beat for the heart is the task of the pacemaker tissue in the right atrium. Much like a metronome will continue to click once started, the pacemaker, once generated, will beat for a lifetime. Beginning as tissue in the embryonic **sinus venosus**, which is considered a primitive and embryonic chamber receiving blood from the body, it becomes incorporated into the right atrial wall as development proceeds. The pacemaker was named the **sinoatrial (SA) node** to represent the embryonic origins and the final adult location. Cells of the SA node are modified cardiac muscle and are connected to the muscle cells of the atrium via the gaps junctions of the intercalated disks. Therefore, when the cells of the SA node spontaneously generate an action potential, it will spread throughout all the cells of both atria through these junctions. However, the muscle of the atria and the muscles of the ventricles are separated by a connective tissue ring called the **annulus fibrosus** that forms the foundational anchor for the valves and the **septa**, or the dividing walls, of the heart. Since the SA node signal cannot spontaneously spread to the ventricles, additional conductive cells pick up and relay the electrical signal through the annulus downward to the ventricles.

In the right atrioventricular region, not far from the tricuspid valve, is another area of modified cardiac muscle called the **atrioventricular (AV) node**. These cells are connected to the atrial muscle via gaps junctions so they are stimulated by the spreading electrical signal from the SA node. In addition to detecting the signal, the AV node also will "pause" before relaying the signal to the ventricles. This will allow the atria to contract just before the ventricles contract and therefore enable the ventricles to fully fill with blood before the next contraction.

Once the pause is completed, AV node cells relay the signal through modified muscle cells that are arranged into a bundle of fibers that run through the annulus fibrosus and down toward the apex of the ventricles. This **AV bundle** (bundle of His) will transfer the electrical signal to the base of the heart, where the fibers spread throughout the ventricle and regenerate the electrical signal in the ventricular muscle leading to the contraction.

## EKG

The health and state of cardiac function can be assessed indirectly and noninvasively by detecting the electrical changes that occur in the heart through the skin of the torso. Electrodes positioned on the chest can receive the electrical potentials and display them as a series of peaks or "waves" that correspond to the electrical activity in the different chambers of the heart. This display is termed an electrocardiogram (EKG).

EKG is used to designate an electrocardiogram because, in German, the heart is referred to as *Kardia*. Also, this convention helps prevent confusing EKG with other diagnostic tests such as an echocardiogram (ECG) or an electroencephalogram (EEG).

The initial small peak seen at the start of a cardiac cycle is the **P wave**, and represents the depolarization of the atrial muscle that precedes the contraction of the atria. This is such a small wave in height because there is so much less tissue in the walls of the atria when compared to that of the ventricles. The next wave is the largest and sharpest and the one more recognizable as an EKG. This is the **QRS wave** (**QRS complex**) for the bottom starting point (Q), the top of the spike (R), and the bottom point (S) of the wave that occurs during ventricular depolarization. Following the QRS wave is an intermediate-sized wave, the **T wave**, and it occurs during repolarization of both the ventricles and the atria.

Any change in the size of the waves or the timing between peaks has diagnostic value for the cardiologist and can be an indicator of a number of cardiac anomalies and illustrate damage that may have occurred.

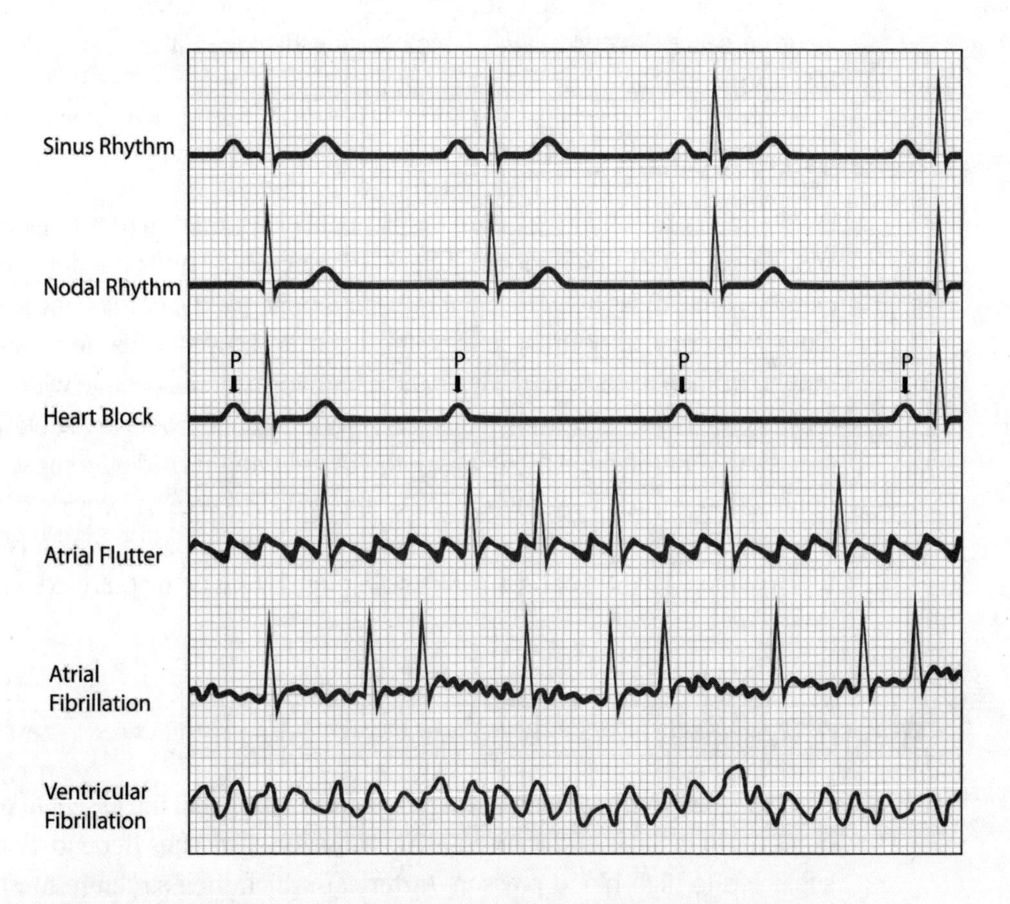


The electrical activity of the heart can be detected and displayed for diagnostic purposes as an EKG. This illustration shows EKGs from normal individuals (sinus rhythm), as well as those with cardiac anomalies such as atrial and ventricular fibrillation.

## Heart Rate

Although the heart is capable of beating on its own, it does not possess the ability to know when to increase or decrease its rate based on activity level. This is where the autonomic nervous system plays a critical role. During heavy exercise, your heart must increase its rate to provide the needed materials including oxygen to the hard-working muscles. This is facilitated by neurons that produce and secrete norepinephrine and the adrenal gland that produces epinephrine (adrenaline). These molecules will cause the cells of the pacemaker to increase the rate of firing and thereby lead to an increase in overall heart rate. Conversely, during periods of inactivity, such as sleeping, other neurons secrete acetylcholine, which causes the pacemaker to slow down and decreases the rate.

## Strength of Contraction

In addition to beating faster, the heart may contract harder to eject more blood per beat during periods of increased demand. This is not dependent on outside signals as is heart rate, but rather it is built into the cardiac muscle itself. During periods of normal activity, the actin and myosin filaments are overlapped in such a way that not every actin can form a cross bridge with myosin. Only as the heart increasingly fills with blood and the fibers are stretched farther apart can the actin fully engage the myosin and provide the most intense contraction. In this way, the design of the muscle proteins provide a built-in reserve to be used when the heart is filled with more blood, which is during times of increased demand and results in more blood being pumped out of the heart.

# Blood Vessels

Blood vessels permeate the human body and convey all the essential components, both cellular and molecular, upon which all tissues depend. Some are subjected to high blood pressure (**arteries**), which their structure must resist, while others (**veins**) must assist the extremely low-pressure blood in returning to the heart. Between these two are vessels that are only 1 red blood cell in diameter and that allow the exchange of materials between the blood and the body's tissues. The functions, similarities, and unique characteristics of these vessels will be described in this section.

Connected end to end, all of the blood vessels in the body would extend approximately 60,000 miles. For reference, the circumference of the earth at the equator is just under 25,000 miles.

## Arteries

The vessels that transport blood away from the heart are called arteries. Although most arteries will be rich in oxygen (thus colored red in all biology texts), the oxygen content of the blood is not what is used to define these vessels. For instance, the pulmonary arteries transport oxygen-depleted blood from the right ventricle to the lungs. These are low in oxygen, but since they move blood away from the heart, they are classified as arteries. Arteries, like all types of vessels, have a lining layer of epithelium called the **endothelial layer**. These cells, because of their overall net negative surface charge and molecular composition, will provide a low-friction surface for blood cells and platelets and allow for the smooth laminar flow of the blood.

For arteries, this **endothelium** (the name of the epithelial lining of vessels) and the underlying connective tissue is grouped together into a layer termed the **tunica intima**. This is the innermost of 3 layers present in most arteries and veins. The middle layer, which is most prominent in arteries, is the **tunica media** and consists of varying numbers of smooth muscle cells and sheets of elastic fibers (elastic lamina). Together the elastic component (which rebounds following high pressure) and the smooth muscle cells (which control the narrowing or expansion of the vessel diameter) assist in controlling both blood pressure and the flow of blood through different vessels. The outermost layer of the vessels is the **tunica adventitia** and is composed of the connective tissue layer surrounding the vessel.

The largest-diameter arteries, such as the **aorta**, are defined as elastic arteries due to the large number of elastic layers present in the tunica media. Pressure is highest when the blood immediately leaves the heart, and these vessels require the elastic fibers to resist and rebound to the high-pressure stretching.

Most of the intermediate-sized vessels have layers of smooth muscle in the tunica media that may reach up to 40–50 layers thick. These are the muscular arteries and are represented by most of the named arteries in the body. As blood flows farther away from the heart, the layers of the arterial wall become fewer and the diameter of the vessel decreases until all that remains of the tunica media is 1 or 2 smooth muscle cells along with the tunica media. These are called **arterioles**, which are small-diameter arterial vessels, and immediately precede **capillaries**, the thinnest-walled blood vessels and the only ones that allow for the exchange of materials with the cells of the body. Often the smooth muscle of these vessels acts as a pressure regulator to prevent the blood pressure from being too high before entering the capillary network.

## Capillaries

The capillaries are the small vessels that will allow the exchange of materials, such as oxygen and carbon dioxide. These are the smallest-diameter vessels that allow the red blood cells to pass through in single file. They are composed of only a single endothelial layer with little, if any, connective tissue. Therefore, materials only have to move through or between the endothelial cells to enter or leave the circulatory system.

Most of the capillaries in the body are categorized as **continuous capillaries**. The endothelial cells of these capillaries form tight complexes between neighboring endothelial cells and only material passed through the endothelial cell will be transported. There is no place in the body where this is more evident than in the brain, where the continuous capillaries are a part of the blood-brain barrier. In other areas, material needs to be transported more rapidly and the specificity isn't critical. The endothelial cells in areas such as in the kidney have small pores that allow larger material to be transported more quickly. These openings, called **fenestrae**, may be open or covered with a thin membrane to further specify the passage of material. In either case, these capillaries are called **fenestrated capillaries**. The final type of capillary is somewhat like Swiss cheese in that there are almost more holes than cells in these vessels. In organs such as the liver, in which cells are in contact with much of the plasma component of the blood, only the cells are restricted from access to open areas, called sinuses or sinusoids. All other material can pass through these large holes. Thus, these are termed **sinusoidal capillaries**.

## Veins

By the time the blood passes through the capillaries, the blood pressure has been drastically decreased. In the aorta, the blood pressure was approximately 100 mmHg (millimeter of mercury). When the blood finally returns to the heart, the pressure is almost 0 mmHg. The exchange of materials in the capillaries and the capillaries' small diameter allow this low-pressure blood to enter the venous and begin the return pathway back to the heart.

The first vessels of the venous system are called **venules**, and are primarily a single endothelial layer. However, their diameter is larger than that of capillaries. Additionally, they usually are found adjacent to an arteriole. From the venules, the blood will move into increasingly larger- diameter veins. This increase in diameter causes the blood to pool and also decreases the pressure of the blood, making it more difficult to move back to the heart, especially for blood in the feet or lower legs.

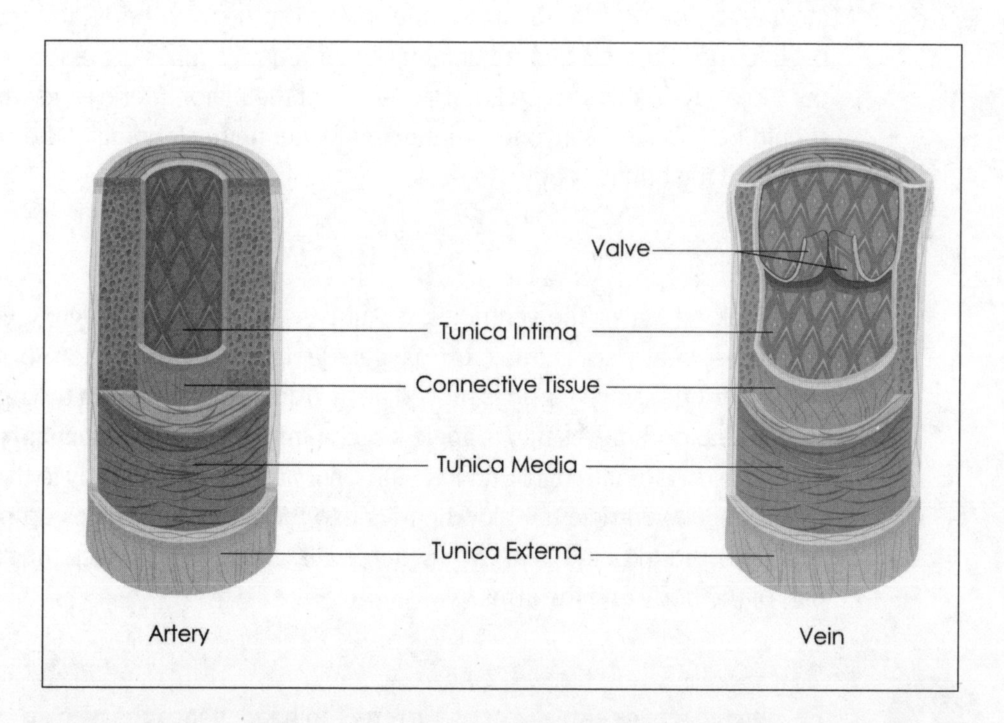


Arteries and veins, while composed of the same basic layers, differ in the thickness of the walls (arteries have thicker walls than veins) and in their diameter, Veins have a much larger diameter when compared to the thickness of the vessel wall.

Unlike arteries, the thickest layer of veins is the **adventitia**, or the outer connective tissue layer. Little smooth muscle is present in the tunica media of veins. Another identifying characteristic of veins is that their diameter will be much greater than the thickness of the vessel wall (the opposite is true for arteries). Since the blood lacks sufficient pressure alone to return to the heart, veins possess unidirectional valves, resembling in some ways the action of the semilunar valves. With each contraction of the heart, blood is moved upward through a set of valves, which will then close as the heart relaxes and the blood attempts to fall backward due to gravity. In this way, the blood is moved upward against gravity through a series of valves, much like ships are moved through the Panama Canal through a series of locks.

## Circulation

Moving the blood from heart to lungs and from heart to body and returning back to the heart is a monumental task that requires miles of vessels throughout the body. In this section is a list of some of the major arteries and veins that should be familiar to anyone looking for a better understanding of the vascular system of the human body.

### Major Arteries

As blood leaves the ventricles, it will pass into one of the "great vessels." While the great vessels transport oxygen-depleted blood, they are both considered arteries since blood is moved away from the heart. From the right ventricle, the blood moves into a single large pulmonary trunk (**pulmonary aorta**) and soon divides into right and left pulmonary arteries on the way to the lungs. From the left ventricle the blood enters into the aorta that arches up over the superior portion of the heart, where large vessels leave to service the superior part of the body and the arms.

*Brachio* means arm and *cephalic* refers to head; hence the naming of the first branch off the aorta, which goes to the head and arm on the right side.

The first branch from the aortic arch is known as the **brachiocephalic**, which will send blood to the right side of the body, including head and neck via the right common carotid artery and right arm. This occurs through the right subclavian artery, which extends through the body wall and into the arm as the brachial artery.

The left common **carotid** and the left **subclavian artery** branch both come directly from the aortic arch and serve the left side in the same manner as the vessels on the right. The aorta continues to arch 180° and descends into the body as the **systemic** or **descending aorta**, from which branches to the remainder of the body will originate. The largest branch off the aorta is the **celiac artery**, which supplies most organs of the upper digestive tract. From the celiac, many arteries will branch out to go to the liver (**hepatic artery**), the stomach (**gastric artery**), and other organs such as the spleen (**splenic artery**). Renal arteries will branch directly from the aorta and deliver blood to the right and the left kidneys. Moving farther down, first the **superior mesenteric** and the **inferior mesenteric arteries** will sprout from the aorta and supply portions of the small and large intestines, respectively.

At the most inferior end, the aorta will divide into the right and left common **iliac arteries** that descend into each leg. This is in much the same way as the common carotid ascends into the head and neck before branching off. The iliac arteries then become the external iliac arteries in the pelvic region, and later the femoral artery as it continues down the leg to supply the muscles of the leg and the foot.

## Major Veins

Many of the major veins in the body mirror those of their counterparts of the arterial circulation and are named as such. However, some veins have different names. Blood from the feet and legs will be drained from tissue via the **femoral vein** into the common iliac. This will move blood into the **inferior vena cava**, the major venous drainage for the abdominal and inferior part of the body.

While many of the abdominal veins will be named as their arterial counterparts (superior and inferior mesenteric, gastric, splenic, etc.), the majority of the veins from organs of the digestive tract will drain into a large vein called the **hepatic portal vein** and will be transported into the liver.

A portal system is a circulatory pathway in which veins move blood into a capillary network and back into veins.

In the liver, blood will pass through the **sinusoidal capillaries** to the liver cells in order to be metabolized. Here the material can be processed, excreted, or stored. From these sinusoids, blood is moved out of the liver by the hepatic vein and into the inferior vena cava.

Blood from the head and neck is drained by large veins on the right and left side called the internal and external jugular veins, which empty their blood into the arching right or left brachiocephalic vein. These join together in the midline of the body to form the **superior vena cava**, which returns blood to the right atrium. From the arms, blood moves upward into the brachial vein, which becomes the subclavian vein once inside the torso at approximately the shoulder region and joins onto the brachiocephalic to return to the heart.

## Blood Pressure

Hydrostatic pressure (blood pressure) is the pressure the blood exerts on the vessel walls due to the contraction of the ventricles of the heart. With each ventricular contraction (systole), high-pressure blood is ejected into vessels of the body. For the average person, the pressure in the brachial artery during this contraction will be approximately 120 mmHg, and is termed the systolic pressure. During a ventricular relaxation (diastole), when the ventricle is expanding and creating a lower-pressure area to pull blood from the atria, the pressure will fall to near 0 mmHg inside the ventricles. However, because of the semilunar valves in the great vessels and the elastic nature of these arteries, the blood pressure will not fall below 80 mmHg on average, and represents the person's diastolic pressure.

**Korotkoff sounds** are the sounds made during the determination of blood pressure using a blood pressure cuff and stethoscope.

Typically these measurements are taken by restricting blood flow through the brachial artery in the arm and then listening to the blood as it returns to normal flow. Using a blood pressure cuff (sphygmomanometer), the pressure on the brachial artery is increased beyond that of the person's systolic pressure, and this blood cannot flow through the artery. At this point, silence is heard when a stethoscope is used because no blood is flowing. As the pressure is released, blood will push through the blockage as soon as the blood pressure exceeds the pressure of the cuff. This is when the systolic pressure is recorded. This motion can be heard as a thumping sound as the blood spurts through with every contraction. As the pressure continues to be released, different sounds will be heard as the blood flows past the blockage in a turbulent fashion, creating the noise. Finally, when the pressure of the cuff is at or below that of the diastolic pressure, the blood returns to smooth laminar flow and is silent. The pressure of the cuff following the final sound before silence will mark the diastolic pressure.

## Diseases and Disorders

Any problem that causes the pumping action of the heart to be diminished or restricts blood flow into specific areas of the body can lead to reduced life expectancy or imminent danger of death. One major problem facing many populations, especially in developed countries, is that of **ischemic disease**, or lack of oxygen to tissues due to poor circulation.

### Ischemia

Because of diet or heredity, many people have high cholesterol levels in their blood streams. Over the long term if left untreated, this material can accumulate in the walls of blood vessels and narrow the passageway of the blood. In some arteries, this may not cause an immediate problem; however, if the narrowed vessel is in the heart, lungs, or brain, the danger may be grave. Some materials in the blood will help to remove these cholesterol accumulations in the blood vessel, while others will actually make this situation worse and lead to greater deposits. Often referred to as "good" and "bad" cholesterol, these lipoproteins can be identified by your doctor, and your lifestyle can be adjusted based on your cholesterol report.

High-density lipoprotein (HDL) is the lipoprotein referred to as the good cholesterol, while low-density lipoprotein (LDL) is called the bad cholesterol.

## Myocardial Infarct

If an ischemic event occurs in the vessels supplying the heart itself, the cardiac muscle, which is not designed to function under hypoxic (low-oxygen) conditions, will be destroyed. This can lead to a heart attack. A heart attack is actually muscular death and degeneration of the heart muscle. It will not regenerate, but will be replaced with connective tissue. As one can imagine, if enough muscle dies, the heart will either function poorly or not function at all. One such vessel that is critical to remaining open is the left anterior descending (LAD) coronary artery. It supplies two-thirds of the left ventricular muscle, and if it becomes blocked, a person will suffer a "massive" heart attack, from which survival is unlikely.

## CHAPTER 11

# Blood

While the cardiovascular system may be thought of as the interstate highway system of the body, the blood and its many components are most definitely the vehicles, carriers, transporters, and workers that keep the body supplied with essential materials. The blood volume of an average adult is approximately 5.5 liters (L), and will contain millions of cells and other molecular components. Additionally, some of these components act as garbage trucks, which remove wastes so that living areas can remain clean and healthy. In this section, the components of the blood will be identified, their functions described, and their roles in body defense and repair explained.

## Plasma

When most people think of blood, their first thought is of red blood cells. In fact, these cells and other formed elements of the blood make up a smaller percentage of total blood volume when compared to the liquid portion of blood: the **plasma**. For an average adult, the plasma comprises 55 percent of the total blood volume, and as with the rest of the body, it is primarily composed of water. In addition, the plasma contains dissolved gases, charged ions such as sodium and potassium, fats, carbohydrates, vitamins, minerals, and proteins.

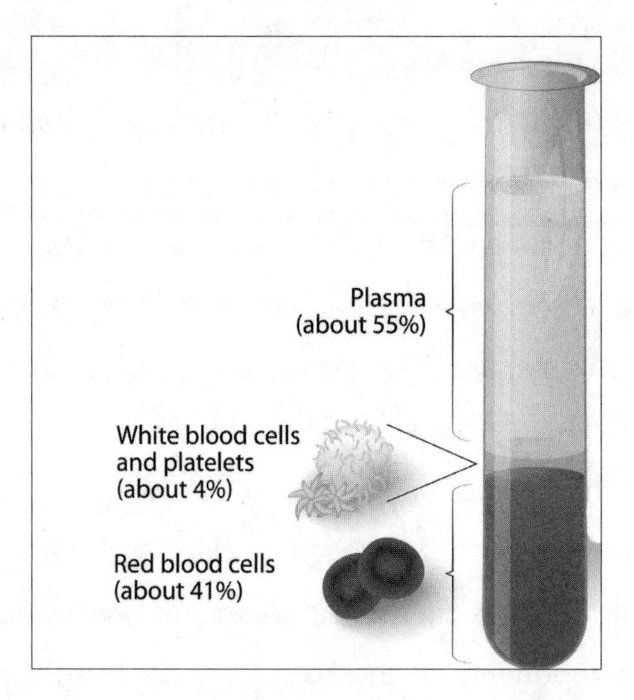


Blood elements have been separated by density to show the proportion of the blood that is red blood cells, white blood cells, and platelets.

## Albumin

The most abundant materials dissolved in plasma are the plasma proteins. Albumin, which is produced in the liver, makes up the bulk of plasma proteins and serves primarily as the "stuff" in the plasma. Albumin acts as a solute in the blood. In this way, albumin ensures that more protein (stuff) will be present in the blood stream when compared to the outside of the blood vessel. This creates an osmotic pressure imbalance that will cause water to be drawn out of the tissue of the body and into the blood. When proteins, rather than ions, salts, or other materials, function as solutes, the water-drawing force is referred to as **colloid osmotic** (oncotic) pressure. Without this vast supply of albumin, much of the fluid that leaks from your capillaries would remain in the tissues of the body and result in severe swelling (edema). Clinical edema can occur when protein production is decreased in cases of liver damage or disease.

Albumin also acts a carrier substance. Many materials that must be transported in the blood stream are insoluble in water, which is a huge obstacle, since blood is mostly water. Because albumin is soluble in water, it can bind to and completely coat other nonsoluble materials. In this way, the entire package appears to be a ball of albumin, when, in fact, the core of the package is water-insoluble material such as bilirubin.

## Globulins

The second most abundant group of proteins in plasma is the **globulins**, which are divided into 3 subcategories labeled **alpha-**, **beta-**, and **gamma-globulins**. The alpha- and beta-globulins are water-soluble proteins also produced in the liver that function as carrier molecules, transporting water-insoluble materials such as lipids and certain vitamins. However, the best-known globulins are the **gamma type**. A more recognizable name for gamma-globulins are **antibodies**, which function as part of the **humoral immune system** to fight infectious agents in the body and provide long-lasting immunity.

## Fibrinogen

Fibrinogen does not provide a transport or carrier function; rather, fibrinogen acts as an emergency repair protein in the event of damage to a blood vessel. In other words, this protein is essential in forming a blood clot and preventing blood loss. While this is the most abundant of the clotting factors in the plasma, there are many other clotting factors that play a role in what will eventually lead to the activation of fibrinogen and the initiation of the blood clot. This will be detailed later in this chapter.

## Red Blood Cells

The oxygen-carrying cells of the blood are aptly named the **red blood cells** (also called erythrocytes and RBCs). These cells give blood its red color. Lacking a nucleus, these cells, once formed, have a limited lifespan, and will circulate through the body thousands of times during their life. Packed with the oxygen-binding molecule **hemoglobin**, these cells shuttle oxygen from the lungs to the tissues of the body where it can be used. Additionally, while only a portion of carbon dioxide produced in the body is transported bound to hemoglobin, RBCs are essential for the processing of $CO_2$ and its transportation in the plasma as bicarbonate.

The average **hematocrit**, which is the volume of blood composed of the cells and platelets, is approximately 45 percent. The red cell count (RCC) is the number of red blood cells contained in 1 ml of blood and averages around 5 million cells for a typical adult male.

### Formation

**Erythropoiesis** is the process of forming RBCs from precursor cells. Early in embryonic development, these cells are clustered together in groups outside of the developing embryo and are delivered to the embryo via a primitive circulatory system. Later in development, the liver will produce RBCs, a task that will eventually be taken over by the marrow of the long bones (myeloid tissue). This is the final location where RBCs are produced in the adult. However, in some instances, the liver and even the spleen may begin to produce red blood cells.

The number of RBCs in the blood remains fairly constant as the newly formed RBCs equal the number of old RBCs that are removed from the circulation daily. This balance is under close hormonal control to ensure that the numbers do not increase or decrease beyond a functional range.

While too few RBCs per unit of blood (**anemia**) is understandable from a functional perspective, having too many RBCs may seem like a good thing in terms of delivering more oxygen. However, too many RBCs (**polycythemia**) will increase the thickness (**viscosity**) of the blood and result in poor circulation and high blood pressure.

When the body detects decreases in the amount of available and transported oxygen (**hypoxia**), the hormone **erythropoietin** (EPO) is released from the kidneys and leads to increased synthesis of RBCs in the bone marrow. These additional cells will aid in the transportation of sufficient oxygen to the tissues when environmental oxygen is low (i.e., at higher altitudes and lower atmospheric pressures). Once the number of RBCs is balanced with available oxygen and oxygen utilization in the body, EPO levels decrease and RBC formation returns to normal levels in a negative feedback manner.

Feedback loops in physiology refer to the body responding to sensory information in order to achieve balance. For negative feedback, the response is the opposite of the sensory input (i.e., if your blood sugar is too high, your body lowers the sugar level). Positive feedback, which is like a chain reaction, is rare in physiological systems, and can be defined as an enhanced response to amplify the sensed stimulus (i.e., uterine contractions during labor will lead to increased production of the hormone oxytocin, which then stimulates the uterus to contract more rapidly and faster, which leads to even higher oxytocin production). This feedback continues to increase until the baby is delivered.

## Structure

As RBCs develop in the bone marrow, the cells progressively become smaller and redder in color as the cytoplasm fills with hemoglobin. Toward the end of their development, round RBC precursor cells called **orthochromatophilic erythroblasts** abandon their nucleus, which allows the cytoskeleton of the RBC (now called a **reticulocyte**) to adopt its typical and final biconcave shape. Since the role of the RBC is to be a bag of hemoglobin for oxygen transport, it may not seem immediately logical that you would want to have any other shape but spherical to pack as much hemoglobin inside the cell as possible. This shape, however, maximizes the number of hemoglobin molecules that will be able to encounter and bind to oxygen. If the cell remained round, with hemoglobin throughout, the molecules at the very center of this round mass of hemoglobin would never encounter oxygen, because they would be too far away from the plasma membrane for diffusion to effectively allow oxygen to reach that depth. Thus, with a biconcave shape, every hemoglobin molecule is within the maximum distance from the membrane and will bind oxygen to a maximum amount.

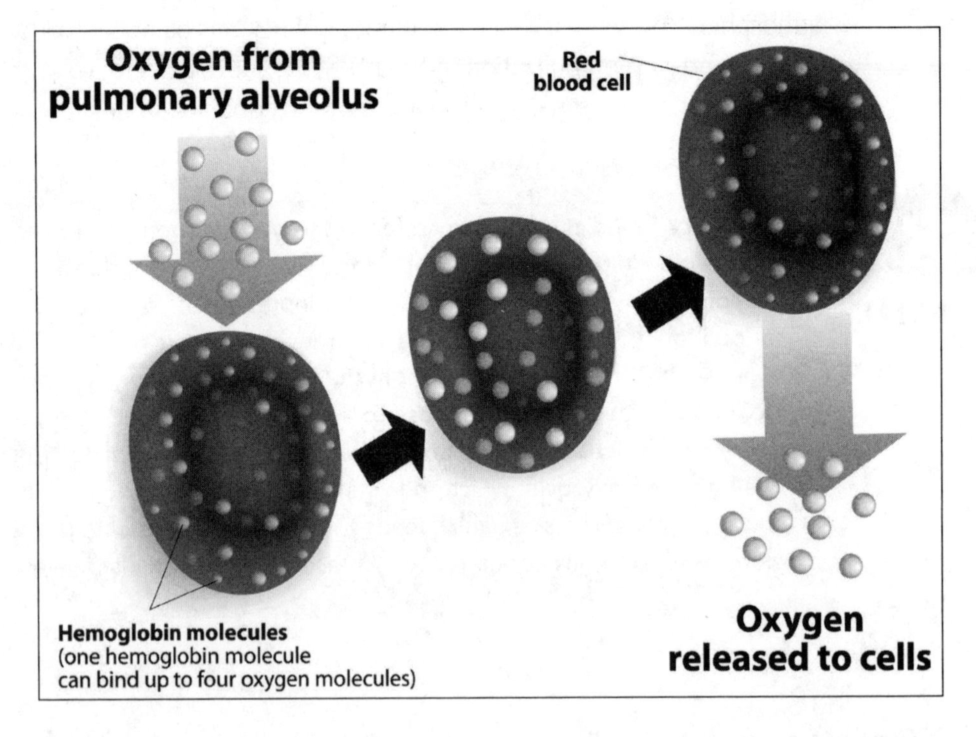


Red blood cells are filled with the oxygen-binding molecule hemoglobin. These cells also are important in the transport of carbon dioxide.

## Hemoglobin

This oxygen-binding molecule is actually 4 protein molecules joined together into a larger composite molecule. Adult hemoglobin is composed of 2 alpha and 2 beta chains, each of which has an amino acid structure called a **heme group** that is capable of binding to a molecule of inorganic iron. It is this iron molecule that actually binds to oxygen in a reversible manner, which is dependent on the partial pressure of oxygen in the immediate environment. Therefore, each hemoglobin molecule binds to 4 molecules of oxygen. When oxygen is bound to hemoglobin (creating **oxyhemoglobin**), RBCs turn a red color. Conversely, when oxygen is unloaded from hemoglobin (creating **deoxyhemoglobin**, the cells turn bluish in color.

## Destruction

As RBCs age and approach their 120-day life expectancy, their plasma membrane becomes more rigid and the cell as a whole is less flexible. Recall that a capillary is only about the diameter of a single RBC (8 micrometers). Therefore, RBCs must be flexible enough to squeeze through capillaries in single-file fashion. Old, rigid RBCs run the risk of blocking capillaries and causing clotting. As the cells circulate they will pass through the spleen, where they are cleaned or destroyed. Within the spleen, sinuses are formed from **stave cells**, structural cells of the sinuses that form into barrel-shaped structures with spaces in between for RBCs to squeeze through. Healthy RBCs traverse the spaces of the sinusoids effectively and in the process have debris cleaned from their surface (pitting). Think of it like wet clothes being forced through two rollers to push the water out of the clothes. However, the older, rigid RBCs will be shredded as they are forced through the narrow space and thus destroyed and removed from circulation.

In the spleen there are also an abundance of resident **macrophages**, which are the vacuum cleaners of the body. These phagocytic cells will remove the cellular debris and process the material to be removed or recycled by the body. One important molecular component that must be effectively processed is hemoglobin. The macrophage begins the breakdown by cleaving most of the amino acids of the hemoglobin alpha- and beta-chains. However, the heme group is processed into a molecule called **bilirubin**. This water-insoluble molecule is bound to albumin and transported in the plasma to the liver, where it is further processed and conjugated to glucuronic acid into various byproducts that can be removed in the feces (such as stercobilin, which gives feces the brown color) or in the urine (such as urobilinogen). If the liver is not fully functional, as is the case in many newborn babies or in patients suffering from liver damage, the bilirubin will accumulate in the tissues of the body and give the skin a yellowish color, signifying that the person is suffering from jaundice.

FACT

Jaundice is the yellowing of the skin from the accumulation of bilirubin in the blood and tissues of the body. This can reach toxic levels in the body if not effectively removed. Fortunately, this is easily accomplished by UV light treatments.

## Blood Groups

Marker proteins and carbohydrates on the surface of red blood cells are formed into groups, which allow for identification and matching of blood cells from donor to recipient in clinical and emergency cases. While there are several blood groups, the 2 most clinically significant will be explained here.

### ABO Group

The most familiar of the blood groups, and the one which most people refer to when asked what their "blood type" is, is the **ABO blood group**. Typically, students think of these as antigen markers on the surface of red blood cells, where A and B are antigens (i.e., antibody-generating substances that elicit an immune response culminating in the production of antibodies against the antigen), and O is the absence of a protein on the RBC surface. This is not the case.

All RBCs will have the same foundational protein on their cell surface. This base molecule is called the **H antigen**. If no additional carbohydrate modifications to this H antigen are present, then these cells would be found in the blood of someone with O type blood. This protein does not elicit an immune response, which can cause confusion when students erroneously think that there is no protein marker at all. What changes the H antigen from O type into either A or B type is an additional carbohydrate that may be added to the base

H antigen molecule. A person with A type blood will have an H antigen with an additional n-acetylgalactosamine, while a person with B type blood will have an additional galactose carbohydrate on the base H antigen. Since ABO genetics is a form of codominant inheritance, a person with AB blood type would have some H antigens with the A type carbohydrate, while other H antigens would have the B type carbohydrate. There would never be both carbohydrate antigens on the same H antigen.

While a person with O type blood will never initiate an immune response to the H antigen, there will be circulating antibodies against the A and B type antigens. This has raised an interesting question about how a person can develop antibodies against an antigen his immune system has never encountered. Many have attempted to explain this problem by hypothesizing that either environmental agents similar to the blood groups or viruses that deliver these antigens infect every human in the early months and years of childhood and stimulate the production of these antibodies.

A person with O negative blood is considered the universal donor, while a person with AB positive blood is the universal recipient.

A person with O type blood has antibodies against A and antibodies against B type blood, while a person who has the A and B antigens on her RBCs will have no antibodies in her plasma. This dictates who can donate blood to whom. Someone with O type blood can donate cells to anyone, since there are no antigens on the cell surface. On the other hand, a person with AB blood can receive blood from any type since they have no circulating antibodies.

## Rh Factor

The **rhesus (Rh) blood group** is so named because of the monkeys in which it was first identified. Rather than being a single antigen, several different genes can be expressed (manifested) on the red blood cell surface and result in a person being Rh positive. As mentioned, several genes may result in the person having Rh-positive blood type; however, the most common Rh antigen is RhD.

In fact, the vast majority of the human population is Rh positive. Only when no Rh antigen is on the RBC surface is the blood considered Rh negative.

Another contrasting feature of Rh factor with ABO group antigens is that no antibodies are found in the plasma of a person with Rh-negative blood. Antibodies against Rh factor will only be produced if and when the blood of an Rh-negative person comes into contact with Rh-positive blood. In normal life, this would be a rare event for most people. However, for Rh-negative females who may become pregnant with babies carrying the Rh factor, it could present complications (this will be discussed in the disease section).

## White Blood Cells

The remaining cellular components, which account for less than 1 percent of blood volume, are called the **leukocytes**, or white blood cells (WBCs). While RBCs transport oxygen, the WBCs function as essential players in the immune system and aid in clearing cellular and pathogenic debris from the body.

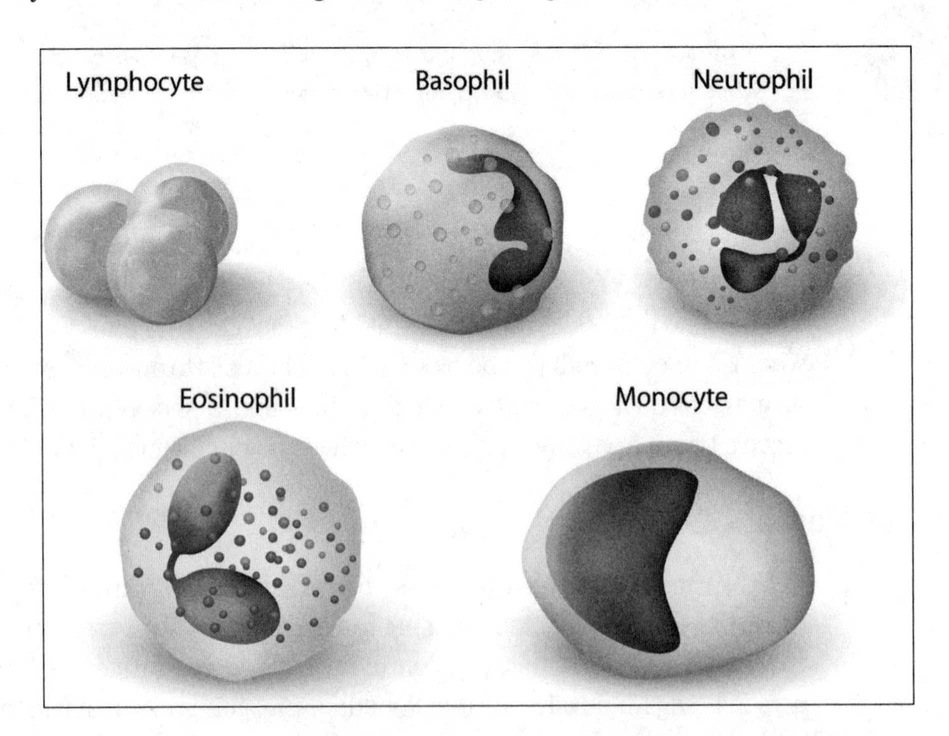


Within the broad category of white blood cells, there are five distinct cells that comprise the white blood cell population in humans.

## Leukopoesis

Stem cells for the WBCs are present in the marrow of bones, much like the precursors for the RBCs. Through a series of developmental stages, the cells will mature and become specified for their particular function. However, many of the WBCs will not fully mature in the bone marrow but do so in other organs of the body instead. Therefore, these cells develop not in the myeloid tissue (bone marrow), but in lymphoid tissue such as the thymus, spleen, and lymph nodes. During this maturation time, the WBCs will distinguish themselves based on the presence (or absence) of specific granules, which gives the cells their color. Those cells with granules are collectively called **granulocytes**, and those without are **agranulocytes**.

The cells in this group are all **phagocytic** cells, meaning these cells will engulf pathogens and debris with their plasma membrane in much the same way that an amoeba feeds. Once the cellular membrane captures this material, additional vesicles will fuse with the captured material and its contents. Some of these added vesicles are called **lysosomes** and contain hydrolytic enzymes that will effectively break down the pathogen or debris contained within the membrane compartment. Additionally, these cells have the ability to follow chemical trails to the area of pathogenic attack via a process called **chemotaxis**. Of course, this means that the cells will leave the circulation and mount an immune attack in the tissues of the body. The process of WBCs squeezing through the endothelial cells of capillaries and gaining access to the tissue compartments is called **diapedesis**.

**Neutrophils** are the most abundant of all WBCs and make up about 60–70 percent of the WBC population in healthy individuals. These cells are slightly larger than RBCs, with abundant granules that stain the cell a color similar to the color of the RBCs. Therefore, this color is said to be neutral. This is how the cells got their name.

These cells live, on average, about twelve hours, and are always the first on the scene of an infection. They are also the first to die as they engulf a pathogen and destroy it along with themselves. Another distinguishing characteristic of the neutrophils is their multilobed nucleus. As the neutrophil ages, more and more lobes will appear on the nucleus, with a maximum reaching approximately 5 lobes. With this difference in appearance, these cells are also often referred to as **polymorphonuclear (PMN) leukocytes**.

**Eosinophils** represent 2.5 percent of all WBCs. These phagocytic cells destroy parasitic organisms and function during allergic reactions. Although they are the same size overall as the neutrophils, these have larger granules and stain bright red with the routine eosin and hematoxylin stain. Often, the granules are so large that they obscure observation of the bilobed nucleus of these cells.

**Basophils** are the final member of the granulocytes and make up less than 1 percent of WBCs in the circulating blood. Of all the granulocytes, they have the largest granules (which happen to stain deep blue). This prevents the direct observation of the nucleus. Basophils are also the only member of the granulocytes that adopt a different name when they leave circulation. When they are in the tissues of the body, these cells are referred to as mast cells. Their granules contain many inflammatory mediators, including **heparin** (which inhibits blood clotting) and **histamine** (which increases vascular permeability). Therefore, when someone suffers from sinus congestion and sneezing, especially due to allergies, the mast cells have released these inflammatory mediators and are therefore the cells responsible for the symptoms.

## Agranulocytes

The remaining WBCs that lack specific granules make up both the largest and the smallest of the WBCs. **Monocytes** are the largest of the WBCs, about 3 times larger than an RBC, and make up about 8–10 percent of the WBC population. These are the vacuum cleaners of the body. Highly phagocytic, they move in and out of the circulation, capture and destroy pathogens, but then also return to areas populated with other WBCs (e.g., lymph nodes, thymus) and present antigens to these immune cells in order to initiate an immune response. Like the basophils, monocytes also change their name when they leave the circulation. In the tissue they are known as **macrophages**, and can have specific names in certain organs. For instance, the resident macrophages in the lungs are dust cells, in the liver they are Kupffer cells, and in the skin they are dendritic cells.

The smallest and the second most abundant of the WBCs are the **lymphocytes**. Lymphocytes are organized into 2 groups. B lymphocytes, which start their maturation in the bone marrow, are the cells responsible for the humoral (antibody-mediated) immunity of the body. T lymphocytes, which mature in the thymus, function in the cell-mediated immune system. Both types of lymphocytes are about the same size as RBCs and are identified as having very little cytoplasm that forms a crescent around a large round nucleus.

# Platelets

You've already heard about plasma and the formed elements of blood. Now it's time to talk about platelets. Although composed also of red and white blood cells, the formed elements of blood also contain platelets. While it would be easier to talk about blood as consisting of the liquid and cellular portion, platelets aren't cells, which leads to the designation of "formed elements," rather than cells.

## Formation

**Platelets** (thrombocytes) are formed in the bone marrow from huge multinucleated cells called **megakaryocytes**. These cells shed fragments of their cytoplasm and membrane as small packages called platelets into the marrow space and eventually into the circulation. These cells continue to shed platelets until all that remains is a membrane-wrapped package of nuclei, which is destroyed by resident macrophages. In times of low platelet numbers, the body will secrete the hormone **thrombopoietin** that leads to the development of new megakaryocytes and more platelets.

## Structure

Platelets are packages of enzymes and materials wrapped in a membrane and are about half the size of the RBCs. The number of platelets in your blood can approach half a million per microliter. These fragments are maintained in an inactive state until such time that a vascular injury occurs and the platelets become activated, secrete their contents and processes on their surface form, and become extremely sticky.

## Platelet Activation

Endothelial cells, which line the blood vessel, help keep the platelets happy and inactive by secreting materials such as **nitric oxide** and **prostacyclin**. However, these cells also enrich the underlying connective tissue with a protein called von Willebrand factor (vWF), which is a potent platelet activator. As long as the endothelium is intact and contiguous, the platelets will never encounter this material and will remain inactive. But when an injury occurs, receptors on the platelet surface bind to connective tissue molecules and vWF, signaling the rapid release of the platelet contents, which will cause vascular constriction (to prevent blood loss) and also activate other platelets in a positive feedback manner. This chain reaction of platelet activation at an injury site will cause the platelets to stick to the wound site, to each other, and to RBCs and WBCs, to form a platelet plug and slow the loss of blood from the vessel. Once activated, the platelet plug is the first in a series of steps that will end in the formation of a blood clot.

## Hemostasis

Hemostasis is the biological ability of the body to minimize and stop blood loss from damaged tissue and blood vessels. The body has several mechanisms in place to halt bleeding quickly in the case of a vascular injury. First, the smooth muscle around damaged blood vessels will intrinsically (automatically) constrict to restrict blood flow and limit blood loss. The platelet plug is the next immediate method of slowing blood loss. However, to stop the bleeding altogether, a clot must be formed.

### Contact Activation (Intrinsic) Pathway

The activation of clotting in response to a small, localized cut is accomplished through a series of enzymatic modifications to clotting factors that are present and continually circulating in the plasma of the blood. When connective tissue materials are exposed to these plasma-clotting factors, an activation complex is organized. The initiation complex relies on several factors, including prekallikrein and inactive factor XII (Hageman factor). When these factors bind to collagen, they will convert prekallikrein to kallikrein, which will activate factor XII (FXII). This starts the cascade of activations with FXII activating FXI, which activates FIX. Once activated, FIX combines with FVIII, phospholipids, and calcium to form a molecular complex that will activate FX, which is the first step in the common pathway for clotting.

### Tissue Factor (Extrinsic) Pathway

Both initiation pathways (intrinsic and extrinsic) will occur simultaneously when an injury occurs in the body. In the extrinsic pathway, when endothelial cells are damaged, FVII leaves the circulation and binds with tissue factor (TF), which is a protein produced by connective tissue cells and forms an activation complex with phospholipids and calcium to activate FX and initiate the common clotting pathway.

## Common Pathway

Whether activated through the intrinsic or extrinsic pathway, FX will activate and combine with FV, phospholipids, and calcium to form the **prothrombinase** complex. As the name implies, this will convert the inactive plasma protein prothrombin into the active enzyme, **thrombin**, which transforms fibrinogen into the active and sticky filamentous molecule **fibrin**. The web of fibrin that begins to form is the foundation of the blood clot as cells, platelets, molecules, and more fibrin coagulate in the site of the clot and stop the bleeding. Although the clot is now formed, it remains somewhat fragile, and the fibrin filaments must be cross-linked for stability by FXIII. This could take as much as 45 minutes to fully stabilize the clot.

Removing calcium is one way blood is prevented from coagulating. EDTA (ethylenediaminetetraacetic acid) and citric acid are 2 common molecules added to blood to inhibit clotting by binding calcium.

As part of the wound-healing process, the clot does not persist forever. Rather than rebuild the entirety of lost tissue, the fibrin web will retract and assist in minimizing the space of the injury and pull the healthy tissue closer together. As the healing progresses, it is important to remove the clot to finish the repair. Circulating plasminogen will become activated over time and plasmin will cut the fibrin network slowly into smaller and smaller pieces until the clot is completely removed and the wound can finish the healing process.

**Warfarin** (Coumadin) was first identified from leeches, which use this substance to prevent blood clotting as the parasite feeds. This works by causing a vitamin K shortage at the cellular level and prevents the formation of a calcium-binding amino acid that is essential for the clotting complexes to activate.

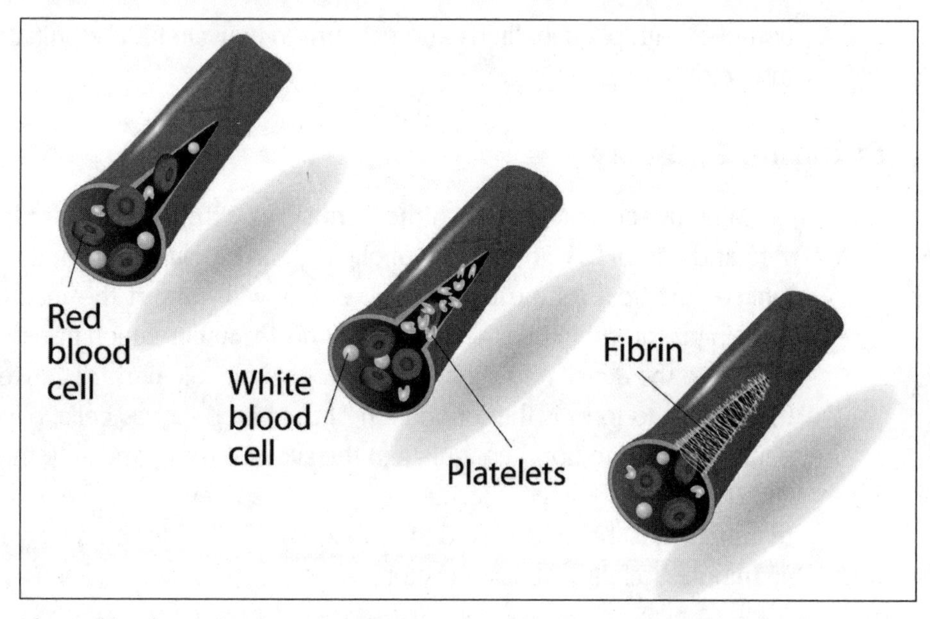


When a blood vessel is damaged, platelets and proteins close the gap. Platelets first form a platelet plug and then proteins start an activation cascade that leads to the formation of a sticky, weblike network of fibrin fibers that will completely close the hole.

# Diseases and Disorders

Since so many components of the blood are essential for so many biological activities, it is not surprising that there is a lot of information about disorders and diseases that affect this system and the blood itself.

## Sickle Cell Anemia

This disease is common in populations of sub-Saharan African descent. Afflicted individuals possess a mutation in the hemoglobin gene that causes the molecule to become rigid and form crystalline structures within the RBC. These abnormal molecules will confer an angular (sickle) shape on the RBCs and cause them to flow less efficiently through the tight passages of the capillaries.

While hemoglobin is formed from 4 protein chains, a person with sickle cell anemia (SCA) will only have a mutation in the beta chains of hemoglobin. In fact, only a single nucleotide in the beta chain gene will be altered. But this is enough to change a glutamic acid into a valine at amino acid 6 and result in the dysfunctional shape of hemoglobin.

The inheritance of this gene stems from its survival in regions of malaria infection. The parasite of malaria reproduces itself within the RBCs of a human host. However, a carrier of this mutation (someone who has only 1 mutant gene and does not have the disease) will have RBCs that, when infected by the parasite, will rupture prematurely, and the parasite will be unable to reproduce in this individual. But as this gene has been passed down through generations that have not been impacted by malaria, the homozygous (having 2 mutated alleles) condition has increased. This has become a health crisis in certain populations.

## Anemia

This condition is most commonly associated with someone having too few RBCs (less than 40 percent hematocrit in most individuals). However, a person may also be anemic if his RBCs do not contain enough hemoglobin. There are a number of causes for anemia that include insufficient iron in the diet, kidney disorders that result in lowered erythropoietin production, and abnormalities of the stem cells in the bone marrow. Another type of anemia, which is termed pernicious anemia, stems from a vitamin $B_{12}$ deficiency. While this may be due to a dietary insufficiency, pernicious anemia is caused by an absence of a factor (intrinsic factor) that is essential for $B_{12}$ to be absorbed by the intestinal epithelium. Intrinsic factor is produced by the cells in the stomach that also produce gastric acid and enables the cells of the small intestine to effectively absorb dietary $B_{12}$. Without intrinsic factor, it is as if the person isn't getting any $B_{12}$ in his diet, and RBC formation declines.

The growing number of people who have gastric bypass surgery where a portion of the stomach is removed may be at risk for pernicious anemia since their level of intrinsic factor will be considerably decreased.

## Thalassemia

While sickle cell anemia is a qualitative defect in hemoglobin, thalassemia results when one or more chains of hemoglobin are not produced in sufficient amounts. Thalassemia is a type of anemia. If the problem is with the beta chain, the classification is beta-thalassemia, or alpha-thalassemia if the deficiency is in the production of the alpha chain.

## Hemolytic Disease of the Newborn

This condition will only occur in situations where an Rh-negative female becomes pregnant and is carrying an Rh-positive baby. If the father is Rh negative, then there is no possibility or concern for this disease occurring. However, if the father is Rh positive, there is either a 100 percent chance of the baby being Rh positive (if the father has 2 Rh alleles) or a 50 percent chance if the father is a carrier and only has 1 Rh allele.

For a first pregnancy, there is no danger to an Rh-positive baby, since the mother does not have pre-existing anti-Rh factor antibodies circulating in her plasma. However, during the birth process, fetal and maternal blood will combine as the fetal portion of the placenta detaches from the uterus and stimulates an immune response in the mother. Antibodies produced against the Rh factor will now be circulating in the plasma, and for a subsequent pregnancy will present significant risk to the baby (if the second baby is also Rh positive). These antibodies will cross the placenta and destroy Rh-positive RBCs in the baby, causing this disorder.

Preventing the mother from mounting an immune response against her baby's blood is a rather easy task. Prior to delivery, the mother is given an injection of anti-Rh factor antibodies. It is done just prior to or during delivery so the baby is not at risk. But the antibodies that will circulate bind to any Rh factor that makes its way into the maternal blood stream. This will effectively eliminate the rogue Rh factor from the blood and render it unable to stimulate an immune response and protect any future babies. This procedure will need to be done with each potential Rh-positive pregnancy.

CHAPTER 12

# Lymphatics

In addition to the arteries and veins, the lymphatic system is composed of a third vessel that transports material throughout the body. While most of the plasma that leaks from capillaries into tissue is returned via those same capillaries, some fluid remains. If this interstitial fluid remained in the tissues and was allowed to further accumulate, swelling (edema) would occur. The lymphatic system is a means by which this leaked fluid may be returned to the circulatory system. Additionally, as this fluid washes through the tissues, it will invariably collect cellular debris and pathogens. To identify and mount an immune response against these materials, the lymph will flow through the vessels and into organs where a host of immune cells are waiting to determine if the materials are pathogens.

# Lymph

The leaked fluid from the circulation is not unlike what is found in the plasma of the blood. However, there are some important compositional and functional differences.

## Composition

Rich in nutrients, electrolytes, gases, and so on, the leaked fluid (lymph) resembles the plasma of the circulatory system. Once in the capillary, high blood pressure will cause plasma to be pushed past the endothelial cells and into the **interstitial tissues**. This, however, will lower the pressure of the remaining fluid in the capillary so that at the end of the capillary (before becoming a venule) the pressure is much lower.

This is where the compositional difference between plasma and lymph becomes important. Aside from cells, protein is the only material that does not leave the circulation. Plasma is rich in proteins, such as albumin, which are too large to leak into the tissues. So, while the blood pressure is lower at the venous end of the capillary, the protein content remains high and continues to exert an osmotic drawing force on the fluid in the tissues. In this way, the proteins are acting as solutes to attract water (there are fewer solutes in the tissues than in the blood) and this attractive force is stronger than the blood pressure trying to push more fluid out. Therefore, the power imbalance favors fluid returning to the capillary. However, some plasma (minus proteins) remains outside the capillary in the tissues where is it called lymph.

## Flow

With little protein content and no hydrostatic pressure (blood pressure) from a pump such as the heart, the lymph has very little pressure allowing it to be pushed or pulled into the lymphatic circulation. The movement of lymph is dependent on pressure exerted in the interstitial tissues by surrounding organs—especially muscle. The action of walking, breathing, or any movement in general will cause the position of organs to change, which will temporarily increase the pressure of the lymph. Inside the lymphatic vessels there is very little pressure, so any external pressure will force the lymph between the endothelial cells that form the walls of lymphatic capillaries and into the lymphatic system.

## Lymphatic Circulation

The lymphatic system begins with narrow, thin-walled, and blind-ended vessels that collect the lymph from tissues and funnel it back toward the heart. Along the path, these vessels will release the lymph into small secondary organs called lymph nodes (covered in detail later in this chapter), which are filled with phagocytic cells (macrophages) and lymphocytes (immune cells). Together, the macrophages will remove and examine the debris, and if any pathogens are found, will present the molecules (antigens) signifying the material as a pathogen to the local lymphocytes in order to stimulate an immune response. After leaving the lymph node, the lymph continues in even larger lymphatic vessels until the fluid, now cleaned of debris and pathogens, is returned to the circulation via the subclavian veins.

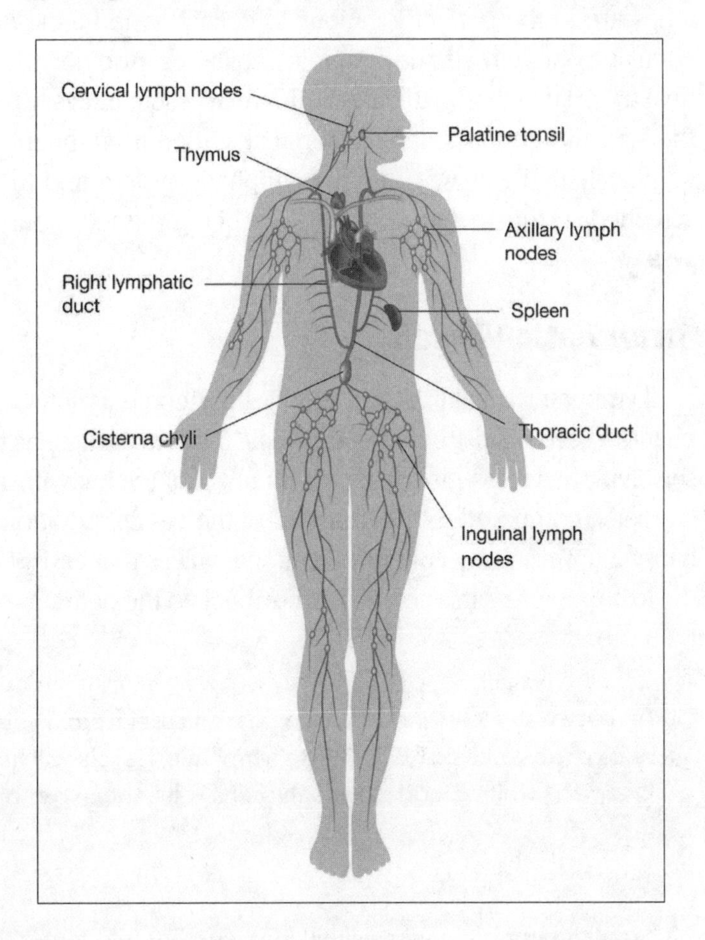


An intricate network of lymphatic vessels course throughout the body in order to return interstitial fluid (lymph) to the circulatory system. Along the path of these vessels, the fluid is filtered and screened by white blood cells in the lymph nodes.

## Lymphatic Capillaries

As mentioned earlier, capillaries are the only vessels that allow for the exchange of materials. Therefore, in order to be removed from tissues, the lymph must flow through vessels called **lymphatic capillaries**. The pressure exerted upon the fluid will force the lymph between the loose and overlapping junctions of the lymphatic capillaries. In essence, these junctions function as unidirectional valves to allow the lymph to come in but not go out. Additionally, the endothelial cells of the lymphatic capillaries are attached to the peripheral connective tissue cells via fibers. These fibers help keep the lumen of the lymphatic capillary open as pressure pushes the fluid into the vessel.

Located within capillary beds of the circulatory system, lymphatic capillaries are in the perfect locations to collect lymph as it fails to return to the circulatory system via the systemic capillaries. Additionally, in the gastrointestinal tract, large lymphatic capillaries are present in every protrusion (villus) of the intestinal surface. Here, material absorbed from the intestines can quickly and easily make its way into the lymphatic system and then be cleaned and screened in the lymph nodes before being passed along to the circulatory system.

## Major Lymphatic Vessels

Lymphatic capillaries will empty their contents into larger-diameter lymphatic vessels. Constructed of the same 3 layers (tunics) as the circulatory vessels, lymphatic vessels most closely resemble veins with their large luminal diameter, compared to the thin wall of the vessel. Additionally, like veins, the lymphatic vessels contain unidirectional valves that assist in the movement of the low-pressure lymph on its journey back to the heart.

An easy way to distinguish a lymphatic vessel from a vein is the presence or absence of RBCs. While lymphatic vessels will have lymphocytes and white blood cells, only veins will contain red blood cells.

Smaller lymphatic vessels will join together into larger-diameter vessels as they make their way back to the heart. Along the path of these larger vessels, you will find lymph nodes, organs rich in lymphocytes and macrophages. Large vessels will combine to form lymphatic trunks, named after the regions they drain. For example, the **lumbar trunk** will drain lymph from the lower abdomen and legs, while the left **jugular trunk** drains the head and neck on the left side. These 5 trunks will all empty into one of 2 **lymphatic ducts**, which will then transfer the drained lymph to one of the 2 subclavian veins and complete its return to the circulatory system. The right lymphatic duct receives lymph from the right side of the body including the arm, thorax, and head and neck. The remainder of the lymphatic trunks of the body will transfer their lymph to the thoracic duct. Therefore, the thoracic is the longest duct and receives the most lymph of the 2 lymphatic ducts.

# Primary Lymphoid Organs

It is important to begin this discussion of **hematopoiesis** (i.e., formation of blood cells) with the location where blood stem cells divide to produce the precursor cells (these will then become blood cells). As explained previously, RBCs begin and mature in the red bone marrow cavities. WBCs begin their development from stem cells in the bone marrow. However, not all WBCs will reach their final mature (differentiated) state in the marrow. Granular leukocytes and macrophages are WBCs that start and mature in the bone marrow. However, the agranular lymphocytes are split between their maturation locations and are named based on that final maturation organ.

## Bone Marrow

While all lymphocytes begin their lives in the bone marrow, only the B lymphocytes will mature there. As naïve B cells, they are not able to recognize antigens with their receptors and are incapable of mounting an immune response. The bone marrow is compartmentalized in such a way that these immature cells are unable to enter the systemic circulation until they pass inspection. If a cell produces a B cell receptor that binds too tightly or too weakly to the antigen, then the B cell is destroyed through one of a few means, the most common being **apoptosis** (programmed cell death or cellular suicide). Additionally, when presented with molecules that are self-antigens, meaning produced by the cells of the host, only those B cells that do not recognize these markers will be allowed out into the circulation. Otherwise, these B cells would initiate an autoimmune response against host cells and tissues.

The B in B cells represents the maturation in the bone marrow, while the T in T cells is because these cells travel to the thymus to complete their maturation process.

## Thymus

Located in the mid thorax (chest) in a location most often referred to as the **mediastinum**, the thymus is an amorphous, bilobed organ that sits superior to the aortic arch and extends upward toward the neck. It is proportionally larger in size during fetal development and into childhood. After puberty and into adulthood, the thymus will curl up and thymic tissue will be replaced with connective and fat tissue. As one of the encapsulated lymphoid organs (i.e., an organ covered entirely by connective tissue), the thymus is divided internally and is isolated immunologically from the body by supportive and structural cells called **reticular cells**.

If you split the thymus in half, you will see that the structure consists of a portion adjacent to the capsule, which is called the **cortex**, and a deeper middle region called the **medulla**. Different types of reticular cells are specific to locations. Here, these cells will isolate one region from another. For example, some will line the capsule and isolate the thymus from the body, while others are found at the boundary between the cortex and medulla to separate these 2 compartments from one another. Additionally, other reticular cells cover the capillaries and blood vessels of the thymus to create a blood-thymus barrier and prevent rogue T lymphocytes from escaping into the body before they are screened.

Naïve T cells begin the screening process in the cortex of the thymus. As they become immunocompetent (i.e., able to recognize pathogens), antigen-presenting cells, such as macrophages and dendritic cells, will determine to what extent the T cell receptor works. As with B cells in bone marrow, binding too tightly or loosely will result in destruction of that T cell. In this way, only adequately binding T cells are allowed to cross into the medulla. This selection is termed positive selection.

Only 20 percent of T cells that start the screening process in the thymus are actually released into the circulation.

In the medulla, functional T cells are challenged with self-markers by other antigen-presenting cells. Only cells that fail to recognize these antigens are allowed to leave the thymus and take up residence in any of the secondary lymphoid organs. This is a negative selection step. The cells that do recognize and bind to self-antigens (and would initiate autoimmune responses) are instructed to undergo apoptosis.

## Secondary Lymphoid Organs

From their primary site of maturation, lymphocytes migrate out into the body and take up residence in local communities of lymphocytes. These cells will accumulate in various locations, such as in organs of the digestive and respiratory tract, as well as in standalone lymphoid organs. Here, they will be better able to quickly and effectively encounter pathogens and mount an immune response.

### Lymphoid Nodules

Often confused with lymph nodes, **nodules** are simply aggregates of lymphocytes. Within the lymph nodes, spleen, tonsils, etc., these groups of lymphocytes appear as dark, dense, rounded areas of tissue that are distinct from the lighter staining and less dense surrounding tissue. Largely composed of B lymphocytes, nodules will also consist of antigen-presenting cells and reticular cells for structure and anchorage. If the nodule is a solid dark color, it is referred to as a primary nodule (follicle). This indicates that the cells of this nodule have yet to be challenged with antigen. After the antigenic challenge, the center of the nodule (germinal center) will become lighter as the lymphocytes proliferate and become antibody-generating plasma cells. These lighter-centered nodules are called **secondary nodules**. T lymphocytes also reside in the nodules but in fewer numbers.

## Lymph Nodes

These encapsulated, bean-shaped lymphoid organs interrupt the path of the larger lymphatic vessels. From the vessels, afferent (conducting inward or toward something) lymphatic vessels will branch out and join at the convex surface of the node. They will then empty lymph into a space beneath the capsule and into the cortex of the lymph node. The cortex is subdivided into compartments by connective tissue extensions of the capsule called **trabeculae**, which help create space for the lymph to flow. It is within this region that the round lymphoid nodules are found.

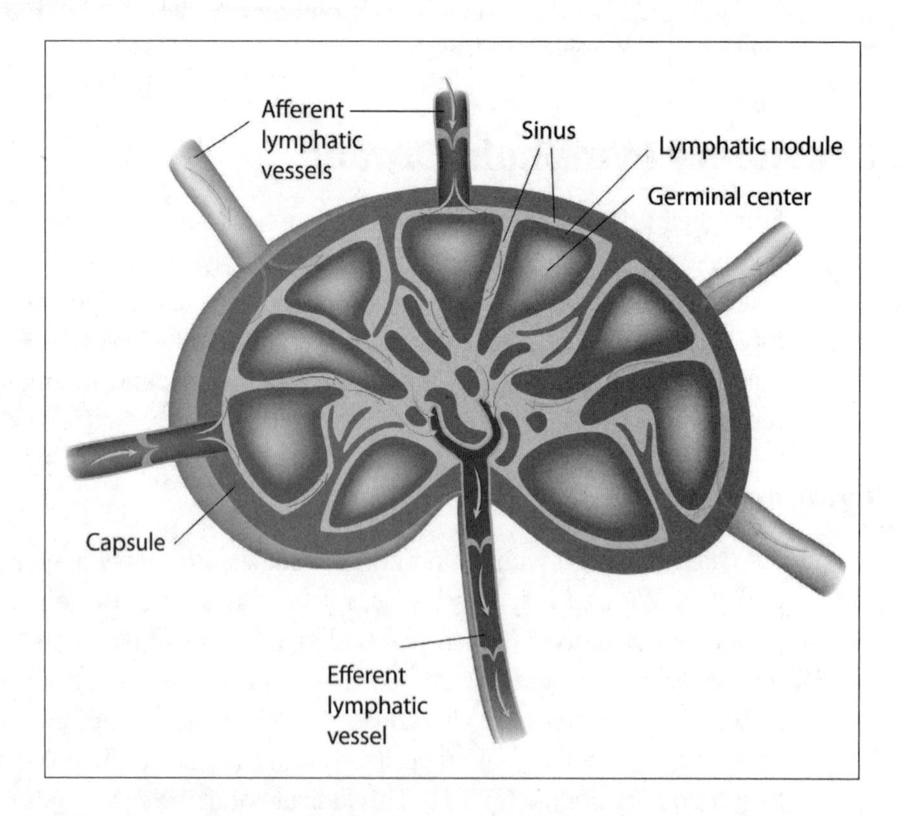


Lymph flows into the lymph node interior, where the debris and materials present in the fluid is screened and filtered by resident white blood cells before the lymph passes out of the node and eventually returns to the circulatory system.

Mostly composed of B cells in these nodules, T lymphocytes are found in the deeper, or **subcortical**, region of the node. Lymph flows through the cortex and down into the medulla (the deeper middle portion) of the node, where spaces or sinuses allow the lymph to collect in the node before flowing into efferent lymphatic vessels, which convey the lymph away from the node.

## Spleen

While the spleen does have lymphoid characteristics, it also functions in the cleaning, destruction, and removal of dead RBCs. Both of these biological roles require the spleen to be highly vascular (i.e., have many blood vessels within). Histologically, these 2 functions are regionalized in the spleen to the red pulp, where RBCs are processed and removed, and the white pulp, where lymphoid nodules and leukocytes are abundant and, therefore, provide immune function.

As the splenic artery enters the spleen, progressively smaller-diameter vessels will branch off and run through the bulk of the spleen. Central arteries are the smaller tributaries that are surrounded by lymphocytes, which form a **periarterial lymphatic sheath (PALS)**. This PALS is further encompassed by a lymphoid nodule, much like that found in the lymph node. Both the PALS and the nodules make up the white pulp of the spleen.

Vessels continue to branch and radiate through the nodules until they form splenic sinuses in the tissue. This is the location of the red pulp. From these sinuses, plasma will freely escape and flow throughout the spleen, including the white pulp where antigen-presenting cells will capture and display antigens to the local lymphocytes. Red blood cells that are nearing the end of their 120-day life span will become brittle, and as the blood pressure forces these cells through narrow spaces between reticular cells that form the walls of the sinuses (stave cells), the RBCs will be shredded. The RBC debris is removed and processed by resident macrophages.

## Tonsils

These organs can become inflamed because of tonsillitis and are often surgically removed. Referred to typically as simply the tonsils, these are in fact one of 3 sets of tonsils in the body and are the **palatine tonsils** (located near the palate). Filled with lymphoid nodules and positioned at the boundary between the oral and pharyngeal cavities (mouth and throat, respectively), these organs are in the perfect location to detect any pathogens trying to gain entry into the body through the oral cavity.

Also protecting the oral cavity, but to a lesser degree, are the **lingual tonsils**. As the name suggests, these are located on the lateral boarders of the tongue and are much smaller in mass than that of the palatine tonsils. These too are filled with lymphoid nodules.

The final set of tonsils is the **pharyngeal tonsils** or better known as the **adenoids**. These are positioned higher in the pharynx at the boundary between the oral and pharyngeal cavities and provide protection from any pathogen seeking to gain entrance to the body.

Histologically, the tonsils are extremely similar. However, to distinguish them histologically: the adenoids are covered with respiratory ciliated epithelium, while the palatine tonsils are covered with oral epithelium lacking cilia.

## Peyer's Patches

While lymphocytes are scattered throughout the body and temporarily accumulate in areas of infection, in the final portion of the small intestine, the **ileum**, permanent lymphoid nodules are present in the **lamina propria** portion of the **mucosa**. These nodules are the distinguishing feature of the ileum, since no other portion of the **alimentary canal** displays such nodules (see terms in Chapter 14 for more details).

## Diffuse Lymphoid Tissue

Although these are not permanent nodules, resident lymphocytes are found in the underlying layers of both the gastrointestinal and the respiratory tracts. Classified as diffused lymphatic tissue, or **mucosa-associated lymphatic tissue** (MALT) as a general term, the lymphatic tissue present in the gastrointestinal tract is known as the **gut-associated lymphatic tissue** (GALT), or in the respiratory tract as **bronchus-associated lymphatic tissue** (BALT).

CHAPTER 13

# Immune System

The world is filled with both beneficial and harmful organisms that only want to survive and reproduce, even if that is to the detriment of a host such as a human. Protecting yourself against macroorganisms and predators is fairly easy, since you live inside a home and typically avoid predatory animals. However, escaping from microorganisms is virtually impossible. Fortunately, your body has evolved a system whereby you can either prevent a pathogen from gaining access to your body or destroy the pathogen before it can do internal damage. This section will look at the many ways your body protects you from bacterial, viral, or parasitic attack.

## Innate and Natural Immunity

While you may think that your immune system is your only defense against dangerous pathogens, there are in fact other tissue and organs in place to protect you from microorganisms while serving additional physiological functions. These innate (nonspecific) means of protecting your body are not about mounting a specific attack against a particular pathogen; rather, these mechanisms are in place to protect you from any pathogenic attack.

| IMMUNE SYSTEM | | | | | | |
|---|---|---|---|---|---|---|
| **Innate (Nonspecific) Immunity** | | | | | **Adaptive (Specific) Immunity** | |
| *Physical/ chemical barriers* | *Fever* | *Inflammation* | *Phagocytosis (cellular eating)* | *Complement* | *Humoral* | *Cellular* |
| Skin–intact barrier<br>Mucus in respiratory tract–physical (sticky) barrier<br>Acid in stomach–chemical barrier | Occurs from the production of one of several factors called endogenous pyrogen | Occurs from the production of one of several inflammatory factors | Action of many of the white blood cells including the neutrophils and macrophages. | Plasma proteins that once activated will lead to the destruction of pathogen, increase the production of inflammatory factors and increase the phagocytic activity of white blood cells | Antibody production | Direct contact or localized secreted effectors (e.g., superoxide ions) |
| | White blood cells | White blood cells | White blood cells | | B lymphocytes | T lymphocytes |

### Physical and Chemical Barriers

The primary physical barrier your body uses to prevent infection is your skin. Your skin is a contiguous layer of cells and provides no space for pathogens to break through. In addition, skin is composed of several cellular layers that pathogens must travel through to gain access to the deeper regions of the body and the circulatory system. If this were not a difficult enough obstacle for pathogens, new skin cells are continually added at the base of the skin and the older cells are progressively moved upward. Therefore, for a pathogen to gain access through intact skin, it would be much like a salmon swimming upstream.

As pathogens try to move deeper into the skin against this upward movement, the top layer of cells is shed daily, so pathogens have a limited amount of time before they are removed along with the dead skin cells. This also brings up an important point about skin. Most viruses require a living cellular host they can latch onto in order to survive and reproduce. The outer layers of skin are in fact dead cells that are compacted together into a watertight barrier. Viruses are unable to reproduce in this barrier before being shed from the surface.

While the skin is an extremely efficient pathogenic barrier, there are other means pathogens use to gain access to your internal tissues. Anything you eat or drink may contain pathogenic agents. Although most foods are processed to destroy harmful microorganisms, the act of touching the food with your hands (especially unwashed hands) may contaminate the food immediately prior to eating. Many of these pathogens, which may be lucky enough to escape the lymphocytes in the tonsils, soon encounter a rather inhospitable environment in the stomach. The hydrochloric acid (HCl) produced to chemically digest our food will also denature and destroy most pathogens that enter the stomach.

Another means of protection relies on trapping pathogens in a thick proteinaceous (protein-filled) substance called **mucus**. Once trapped, the debris- and pathogen-laden mucus can be moved into areas where the pathogens may be eliminated. This is the case in the nasal and respiratory tract. This is where **goblet cells** secrete mucus to trap pathogens, and then **ciliated cells** move the mucus upward to the larynx where the material is transferred to the esophagus and the HCl-filled stomach. Similarly, tears are produced to moisten and lubricate the sclera of the eye. Along with blinking, this mechanism is not unlike using the fluid and wipers to clean the windshield of your car. Tears also contain mucus and can trap any pathogens on the surface, then remove this material from the eye via the **nasolacrimal duct**, which drains into the nasal cavity.

## Phagocytosis and Opsinization

Many white blood cells remove pathogens and debris by phagocytosis. Macrophages are particularly adept at this "cleaning" of the body. These cells remove any foreign material in a nonspecific manner. Additionally, when pathogens are decorated with antibodies and/or are covered with certain complement factors (discussed in the next section), the phagocytic activity of these cells is increased. This elevation of phagocytic activity resulting in more rapid removal of pathogens is called **opsinization**. While an antibody may specifically bind to a pathogen, opsinization doesn't require a specific type of antibody. Opsinization only requires antibodies or complement factors be present and bound to any pathogen.

## Complement

In a cascade, not unlike what happens in the clotting cascade, inactive complement factors in the plasma become activated during a pathogenic attack. The proteins and enzymes that play a role in the complement cascade are simply labeled as C1–C9. The classical activation pathway is initiated via antibody opsinization of a pathogen, which triggers complement factor 1 (C1) to activate C4, then C2. These combine to enzymatically activate C3. These factors are commonly described as having 2 portions: A and B. Thus, C3 is split upon activation into C3a and C3b.

An alternative pathway can lead to the activation of C3 directly when unfamiliar carbohydrates are recognized. These carbohydrates are foreign to the human body, yet rather common in bacterial cell walls.

Once split, each portion of C3 is active. C3a along with another fragment, C5a, will lead to an increase in inflammation in local tissues. C3b, along with antibodies, will further opsinize pathogens. Additionally, C3b will further activate factors C5–C9 to form a protein pore, which will insert into the plasma membrane of pathogens and result in their death. This is referred to as the **attack complex**.

## Cytokines

Considering the function of C3a and how it leads to physiological changes in cell and tissue behavior, many chemicals produced and secreted by immune system cells have far-reaching and powerful effects on the body. Collectively these are referred to as **chemokines**, but similar terms are also often used, such as **cytokines** or **lymphokines**. The vast array of these chemical mediators is beyond the scope of this section; however, the general concept of their actions is essential in understanding immune function. Many of these cytokines play critical roles in the activation of the immune system during an immune response, while others will be required to slow and even halt such a response.

## Inflammation

Several cytokines will affect the blood vessels of the body to allow more blood to flow into the tissue (**vasodilation**) and to make the capillary endothelial cells easier for WBCs to migrate between (increase **vascular permeability**). A consequence of both is to allow more plasma to leak into the interstitial tissues more rapidly than can be removed by lymphatics. At this point, swelling (edema) will occur. Mast cells (basophils that have left the circulation) secrete powerful inflammatory cytokines such as histamine, which makes the vessels leaky. Additionally, mast cells secrete **heparin**, which is a molecule that inhibits the clotting cascade by preventing the activation of thrombin. If the vessels were made to leak and heparin was not produced, a blood clot would rapidly form and prevent the passage of fluid and WBCs into the tissues.

The 4 cardinal signs of inflammation are redness (rubor), heat (calor), swelling (tumor), and pain (dolor).

## Fever

Another effect of some cytokines is to increase the body's temperature above normal. In fact, a group of cytokines is called the **endogenous pyrogens** because of their ability to cause fever. While having a fever isn't a pleasant feeling and everyone wants to reduce a fever quickly, this increased temperature is actually trying to help fight an infection. Higher body temperatures will destroy some pathogens directly, and will reduce the effects of **bacterial endotoxins** in the body. It will also increase the division, migration, and metabolism of the immune system cells and give them an attack advantage over many pathogens. Only when the body temperature reaches or exceeds 105°F will your cells be destroyed. However, any fever lasting for more than a few days and in excess of 101°F–102°F should lead one to seek clinical treatment from a doctor.

Endotoxins and exotoxins are both molecules that lead to the death of human cells when they are in the proximity of bacteria. Endotoxins are present inside of bacteria and released only upon the death of the bacteria. Exotoxins are actively secreted by living bacteria.

## Adaptive Immunity: Humoral

The following sections will address the specific (adaptive) immune system, which is divided into 2 components. The **humoral immune system** relies on B cells that lead to the production of antibody-producing plasma cells. As antibodies are produced, they will link pathogens together into larger and larger masses, which prevents their dispersal throughout the body and provides a larger target for phagocytic cells. Additionally, antibodies will opsinize pathogens for more rapid removal by WBCs.

### Antibodies

When B lymphocytes are activated by their reciprocal pathogen, or the molecule that they will specifically recognize as a pathogen, they begin to rapidly divide and clone themselves into either plasma cells (active antibody-producing cells) or memory B cells. Memory B cells are inactive and held in reserve for a subsequent exposure to the same antigen later in life.

Antibodies are proteins consisting of 4 amino acid polymers that are linked together via disulfide bones between adjacent cysteine amino acids. For a general idea of the shape of an antibody, imagine a person standing with arms upward and legs together making her body into the shape of a Y. This is the overall shape of an antibody monomer. In this analogy to the human body, if you split the body into right and left halves, you would have created the 2 heavy chains that make up an antibody. Two additional chains, the light chains, are smaller proteins that are attached to and parallel to the arms only.

For functional consideration, this antibody monomer will be divided further. If you think of your body and what you use to grasp objects, your hands immediately come to mind. In the previous illustration of an antibody, the hands of the antibody consist of portions of both heavy and light chains. These are globular molecules and the analogy is for general structure, and more specifically, for localization of functional domains of the antibody, which are the variable regions that bind to a specific antigen. While the basic structure of the majority of the antibody molecule remains the same for antibody types, the variable region differs between different antibodies. This variable region enables an almost infinite array of antibodies to protect the body.

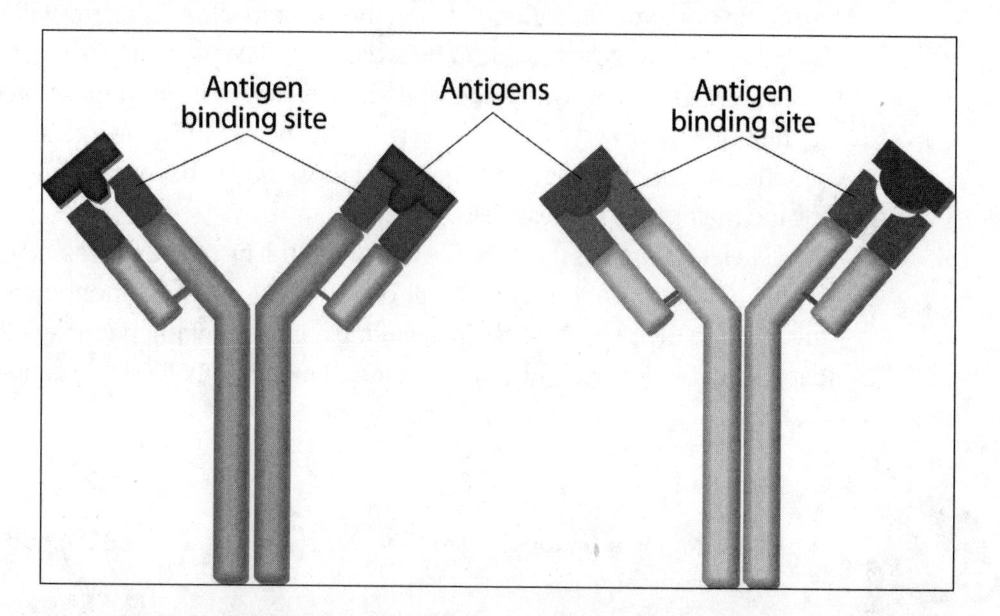


Variable portions of the antibodies allow antibodies to bind specifically to their recognized pathogen.

The remainder of the antibody, which would be the torso and legs, is a constant fragment. These portions are constant in the sense that when antibodies of the same **isotype** have regions that differ, the constant fragments will be identical. These constant fragments, regardless of variable region or isotype, function as an opsin for phagocytic cells.

## Isotypes

While the variable regions of antibodies are specific for their antigen, the constant fragments are the same from antibody to antibody of the same isotype. Antibodies can be produced as one of five isotypes, each having different biological function. The genes for the constant fragment are subdivided into different isotypes that can be used to build an antibody at certain developmental stages or for specific responses. These isotypes are arranged on the chromosome in the order they may be used. The first isotypes used are on the upstream or 5´ portion of the chromosome, while the last isotope will be on the far 3´ end. This is important to remember, since the change from one isotype to another will result in the splicing out of the upstream genes, meaning once you pass an isotype on the chromosome, that gene will be eliminated and no longer available to be switched to later in the life of that cell or its clones. Or, to simplify, when the body makes new cells from those precursors in the bone marrow, you start over with a fresh slate of isotype genes and can produce antibodies from any time, until those genes become eliminated during the maturation of these cells.

The IgM isotype is the most 5´ gene and is the first to be expressed on naïve B cells. This constant fragment will direct 5 IgM antibody monomers to link together via their constant chains, making a large pentameric molecule that is capable of binding ten antigens at a time (remember that a single antibody has 2 variable regions and can bind 2 antigens).

Circulating IgM antibodies are the ones responsible for ABO blood type incompatibility.

The next isotype to be expressed is the **IgD isotype**. This is a monomeric antibody that is also found on naïve B cells and thought to be expressed when the B cells are ready to be challenged by an antigen.

**IgG antibodies** are the predominant antibody found in the circulating plasma. They are also the protein primarily responsible for the humoral immunity. Like the IgD, IgG antibodies are monomeric in structure. Interestingly, these are the only antibodies that can cross the placenta during pregnancy and provide a passive immunity to the developing fetus. However, these isotypes can also cross the placenta and lead to destructive effects, including hemolytic disease of the newborn (erythroblastosis fetalis).

The next isotype is the **IgA isotype**, which is a dimeric antibody, meaning it consists of 2 monomers joined via the constant fragment. These antibodies are particularly resistant to degradation and can be found in many mucus secretions including tears, saliva, and breast milk.

Colostrum (immune milk) is the first secretion from the lactating breasts and contains an abundant supply of IgA antibodies that provide passive immunity to the newborn.

The final isotype is the **IgE** type. This type is capable of eliciting some of the most powerful immune responses of the body. These monomeric antibodies will often find themselves linked to the surface of mast cells via an IgE constant fragment receptor. If an antigen is present, this may result in 2 adjacent antibodies binding simultaneously and bringing the 2 receptors close enough together to trigger a cellular response. In this way, mast cells are signaled to degranulate or rapidly secrete their inflammatory cytokines. When enough of these mast cells release their chemical mediators in a mass explosion, it may trigger a severe to dangerous allergic response and send someone into anaphylactic shock.

## Primary Immune Response

Upon initial exposure of the body to a particular antigen, B lymphocytes, granulocytes, and macrophages stimulate a primary immune response. Leukocytes phagocytose (engulf and destroy) material and produce cytokines that function as signaling molecules to stimulate B lymphocytes to divide. When an antigen binds to the B cell receptor (which is essentially an antibody on the B cell surface), the cell will begin to clone itself into more B cells. Over the course of approximately 2 weeks, plasma cells will produce a low level of IgG antibodies. Initially, the plasma cells that result after B cell activation will produce and secrete the IgM isotype of an antibody, which is characteristic of this primary immune response.

Over the course of the 2-week period, the individual will suffer from the symptoms of the disease as the entire immune system fights to remove the pathogen and prevent further damage. This is the natural primary immune response. However, although a few IgG antibodies result, the principal product of this response is a vast number of memory B cells. These will remain in the body and remain viable for possibly the entire lifetime of the individual. With these in place, when the same antigen is encountered again, there will be a much faster and more robust response.

Vaccines are the clinical way individuals are given the opportunity to mount an immune response, build up their memory cells, and prevent the disease. In many cases, the pathogen introduced artificially will either be already dead or be unable to reproduce, leading to the activation of the immune cells and not the detrimental effects of a living pathogen. In the late 1700s, Edward Jenner introduced this concept by using a less virulent (deadly) strain of cowpox virus to stimulate a primary immune response in individuals. The plasma cells and memory B cells that would result could also recognize and destroy the deadly smallpox virus in his patients.

Because of Jenner's work with smallpox and cowpox, the word "vaccine" is derived from the Latin *vacca*, which means cow.

## Secondary Immune Response

If your body encounters a pathogen, the chances of future encounters are almost a certainty. The purpose of the primary immune response is to have more of the specific B cells and memory cells to quickly remove the pathogen before it can cause harm. While the primary response takes days or even weeks, a subsequent exposure to the pathogen will result in a mass production of IgG antibodies that flood the circulation in a matter of hours. In this manner, the pathogens will be eliminated and the individual will suffer from no symptoms of the disease.

The efficacy of vaccines is often called into question by patients. To ensure long-term protection, subsequent injections are required as "boosters" in order to increase or boost the number of memory cells.

# Adaptive Immunity: Cellular

The second component of the specific immune system involves the T lymphocytes and requires physical contact between cells. It is therefore named **cellular immunity**. There are 4 types of T lymphocytes. Each has specific receptors capable of binding to a reciprocal receptor on cells of the body. Additionally, in much the same way that activation of B cells leads to the production of memory B cells, stimulation of T cells will lead to the formation of memory T cells.

## T Lymphocytes

The first T lymphocyte to consider is the **helper T cell**. These cells are filled with cytokines and are responsible for the chemical stimulation and activation of the immune system. This is true chiefly of T lymphocytes and to a lesser degree B cells and leukocytes. While histologically indistinguishable from any other T cell, helper T cells display a receptor on their surface called CD4, which will be used along with the complementary receptors on antigen-presenting cells to become activated when a pathogen is present. Therefore, helper T cells are often referred to as **CD4 positive T cells**.

**Cytotoxic T cells** express a different CD receptor on their surface, which aids in identification and also in the functional specification of these cells from other T cells. The receptor is CD8 and has a complementary receptor on all cells in the body except for RBCs. In this way, if any cell becomes virally infected or **tumorigenic**, the CD8 receptor will recognize this foreign material and activate a response.

When activated, the cytotoxic T cell will attack the problematic cells with a barrage of cytokines, which will potentially signal the cell to undergo apoptosis. Other chemicals, such as **superoxides** or **perforins**, will directly attack the membrane and proteins of the cell.

Regulatory T cells are also produced during an immune response. While these cells do not directly attach to pathogenic cells, or signal increases in the immune system, they do play a vital role in homeostasis. An immune response is a form of a positive feedback loop: Activated immune cells will activate more immune cells. To halt this chain reaction, regulatory T cells will divide until a threshold mass of cells is produced, at which time they will bind to the helper and cytotoxic T cells and signal them to undergo apoptosis and stop their action. These cells, however, do not affect the memory T cells that are held in reserve.

## Major Histocompatibility Complex Antigens

The complementary receptors for the CD proteins on T cells are the **major histocompatibility complex (MHC)** antigens. The pairings of these molecules are MHC class I, which is on every cell in the body except RBCs and is the binding partner for the CD8 of cytotoxic T cells. If the cell is normal and healthy, the cytotoxic T cell will not be activated. However, if a viral or tumor marker is displayed along with the MHCI, that will lead to T cell activation and cellular destruction.

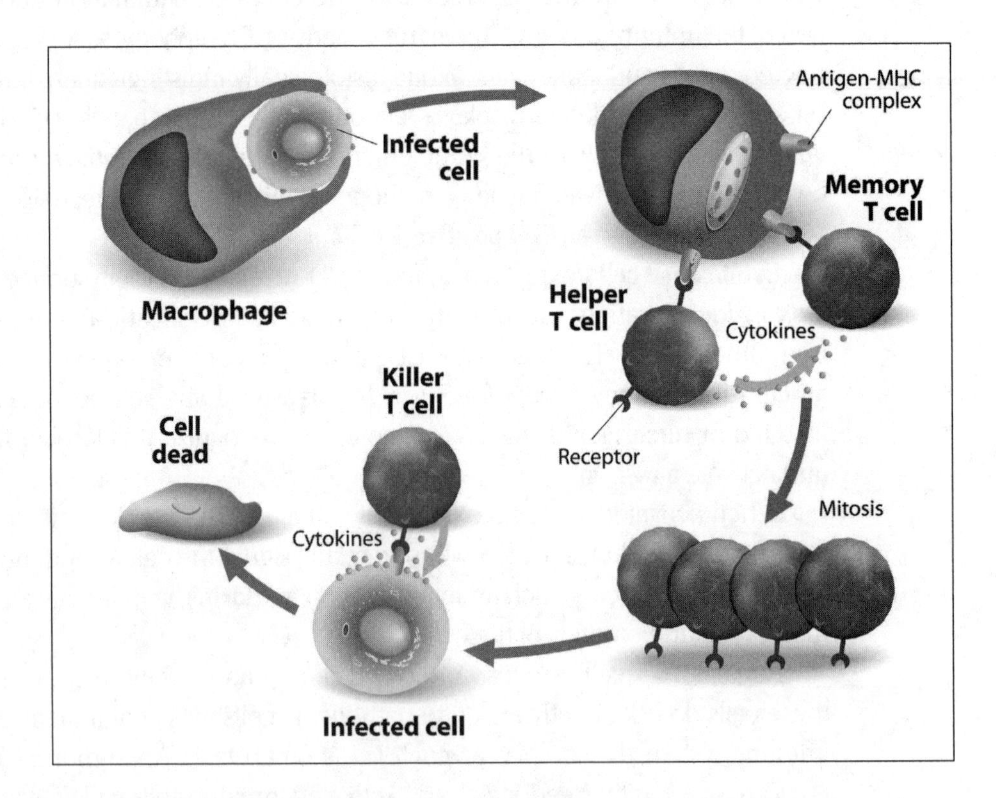


This illustration demonstrates the role of cell surface receptors in the recognition of pathogen and the activation of cell-mediated immunity.

For the CD4 receptor of helper T cells, the reciprocal protein is the MHC class II that is present on antigen-presenting cells, such as macrophages and dendritic cells. These phagocytic cells will destroy the pathogen, display specific antigens on their MHCII molecule, and present this to T cells in order to find the exact match for the pathogen and activate the appropriate response.

## Diseases and Disorders

Given the number of pathogens in the environment, the rearrangement of genes of the immune system, and the screenings that occurs, it is surprising that more doesn't go wrong with the immune system. However, evolution has tested the efficiency of this highly complex system of protection for your body.

### AIDS

Acquired immunodeficiency syndrome (AIDS) occurs when one's immune system functions are impaired. However, the human immunodeficiency virus (HIV) actually takes advantage of the immune system to survive and reproduce. HIV utilizes helper T cells as a host that can replicate and further infect other helper T cells. At first, an afflicted individual will feel symptoms not unlike that of a common cold, which will dissipate in a few weeks. Thinking the disease and danger have passed, the individual will typically not seek clinical intervention and will not display any symptoms for possibly up to 7–10 years. This is the clinically asymptomatic period of HIV infection, when the virus is slowly increasing in numbers as the helper T cells are gradually destroyed. In the end, the immune system is rendered nonfunctional and the individual will succumb to any of a myriad of opportunistic pathogens.

### Allergies

Like AIDS, allergies are not mutations or disruptions of the immune system, but instead are instances of the immune system mounting a response to a common antigen (allergen) that stimulates a massive release of inflammatory cytokines (type I hypersensitivity). This immediate type of allergy is particularly dangerous if the allergen is inhaled or otherwise consumed into the body. The vast numbers of mast cells with their attached IgE molecules will flood the system with histamine and heparin, causing a massive drop in blood pressure and systemic edema (which include the lungs and airways). This is called anaphylactic shock, and if clinical intervention isn't immediately available, death could result in a matter of minutes.

To reverse an anaphylactic response, epinephrine (adrenaline) should be administered. If one suffers from severe type I hypersensitivity, especially to food allergies and insect stings, you must have an EpiPen (epinephrine injector) readily available at all times.

T cells are also involved in allergic responses. These are referred to as delayed (type IV) hypersensitivity. Most commonly these are observed as rashes, hives, or welts that appear on the surface of the skin as a result of contact. One example of delayed hypersensitivity is what occurs when someone is exposed to the uruishoil of poisonous plants. However, this is also the mechanism responsible for tissue transplant rejection if the tissue isn't a close enough match for the patient. The MHC molecules between donor and recipient are being matched at this point. Therefore, siblings (especially identical twins) make the best donors.

## Autoimmune Disorders

Almost daily the list of human autoimmune disorders grows longer. With the numbers of cells produced, invariably and by chance, autoimmune diseases will result. A few of these better known disorders include Crohn's disease of the gastrointestinal tract; lupus, a systemic inflammatory condition; and Hashimoto's thyroiditis, which is the most common cause of goiter in females.

## CHAPTER 14

# Digestive System

The human body is an extremely complex machine. Just like any mechanical device, it requires fuel to power itself. This power is derived from raw material that becomes energy, and this energy is processed in the digestive system. The digestive system can be thought of as a long internal tube (**alimentary canal**). Food enters your mouth and is processed via mechanical and chemical digestion as it is propelled along the meters of your digestive tract. In the final portion of this tube, your intestines stop digesting the material and begin absorbing nutrients into your lymphatic and circulatory systems. The essential materials that your cells and tissues rely on for survival are then distributed throughout the body.

# Mouth

The oral cavity (mouth) is defined as the space bounded by your lips (anteriorly), cheeks (laterally), and the palate (superiorly). In this space, food will be processed by mechanical means (teeth), chemically modified (saliva), and moved into the esophagus by the tongue. Raw materials begin this journey by first passing through the opening (stomodeum), defined by the lips.

## Teeth

Made up of the hardest substance in the body and similar in many respects to bone, the teeth are the cutting and grinding implements that begin the mechanical breakdown of ingested food. All teeth consist of a portion that rises above the gum line (**gingiva**), which is called the crown, and the portion below the gum line (root). Covering the crown of the tooth is a brilliant white, calcium-rich material called enamel. Unlike bone, enamel has no cells and will not repair or produce new enamel. In time, enamel will slowly wear away or be degraded by bacterial enzymes (dental caries or cavities). Beneath the enamel is the core of the tooth, which is composed of **dentin**. This living tissue is similar to, yet harder than, bone and surrounds the pulp cavity where blood vessels and nerves reside. Blood will be transported in and out of the pulp cavity via vessels that enter and leave through an opening in the root of the tooth.

### Primary Teeth

The first teeth for babies will erupt through the gums starting around 6 months of age and may continue erupting for the first few years of life. During this time, the first set of teeth (primary teeth) will form. For each jaw, the teeth will form into pairs in this order, from the front of the mouth to the back: central incisor, lateral incisor, canine, first premolar, and second premolar. This will complete the first set of teeth (5 pairs per jaw, 20 teeth in total). These are also referred to as the deciduous teeth since they will be lost as the secondary (adult) teeth push upward to replace them.

### Secondary Teeth

The final set of teeth, which must last a lifetime, begin to erupt through the gums at around 6 years of age. They may not finish erupting until a person's early 20s. Structurally, both sets of teeth are the same, although the adult teeth are larger to better facilitate chewing (**mastication**) in the larger adult mouth. These teeth typically erupt in the same order as the primary teeth. In addition to these teeth, the adult mouth contains a pair of first, second, and third molars for each jaw. The final total of secondary teeth in the adult mouth will be 32.

**Why are the third molars called wisdom teeth?**
Because they are the last to erupt in early adulthood, after the adult has matured and gained a level of experience (wisdom).

## Tongue

The tongue plays an essential role in digestion. As the jaw muscles and teeth cut, tear, and grind the food, the muscular tongue will move the food back and forth in the mouth (**intraoral transport**) to process the food into small pieces. Additionally, as saliva is added to the mixture, the teeth and tongue will mix smaller and smaller pieces of food with this enzymatic secretion. Lastly, the tongue will create an anterior barrier, which, along with muscular contractions in the pharynx, will move food into the esophagus.

Made up of skeletal muscle covered with a thick, tough epithelium, the tongue is a versatile organ. Individual muscles are arranged into at least 5 different planes within the tongue, which facilitates movement in a myriad of directions. This is certainly evident if you have ever observed an individual folding the tip of his tongue into a clover, which is just a small example of how mobile the tongue really can be.

Raised extensions of the epithelium can be found on the upper surface of the tongue (**dorsum of the tongue**) and will come in contact with the food. Some of these extensions have thickened layers of dead cells making them very tough and abrasive (**filiform papillae**). For example, the rough texture of a cat's tongue is populated with an abundance of these papillae, which are used to move food as well as assist in pulling material into the mouth. Other papillae are more flattened, often with grooves on the lateral boundaries, which can accumulate material and assist in the detection of taste via taste buds that are embedded in the walls of the papillae.

## Salivary Glands

As the teeth and tongue facilitate the mechanical digestion of ingested material, an enzymatic mixture called saliva is added through ducts from 3 major salivary glands. This substance will begin the process of chemical digestion, as well as lubricate the material for passage down the esophagus.

### Parenchyma

The secretory portion of salivary glands is called the **parenchyma** of the gland and will typically consist of either or both **serous** (watery, enzymatic) or **mucous** (viscous, slippery) secretions. Histological identification of salivary gland tissue is largely dependent upon the percentage of each type that composes the gland.

### Parotid

The **parotid** is located on the lateral portions of the face near the angle of the jaw and the base of the ear. The **parenchyma** (i.e., the secretory portion) of the parotid gland is entirely serous in nature and will include enzymes such as salivary amylase that begin the breakdown of carbohydrates. These secretions are transferred to the mouth via a tube called **Stensen's duct** and empty into the oral cavity near the second molar on each side of the maxilla (upper jaw). Although it is the largest of the salivary glands, the parotid does not produce the bulk of saliva.

### Submandibular

As the name implies, these glands are found just inside the lower jaws and produce around 60 percent of all saliva. Unlike the parotid, the submandibular consists of 50:50 serous and mucus-producing cells. **Wharton's duct** transfers the saliva from the gland to the oral cavity, where it is released at swellings on either side of the **frenulum** (i.e., the string-like tissue connecting the ventral surface of the tongue to the floor of the mouth) of the tongue called the **lingual caruncle**.

### Sublingual

The smallest of the salivary glands is the **sublingual gland**. This gland is composed almost entirely of mucus-producing cells. This material will be added to the serous secretions from the other glands and aid in the lubrication of the food before swallowing.

**What structure swells and is an indicator that a patient has the mumps?**

With the mumps infection, the parotid gland becomes greatly enlarged and is the classic symptom that the individual has the mumps.

## General Structure of Alimentary Canal

While the different regions of the alimentary canal may vary in their specific structure and function, each area will have the same 4-layer foundation.

### Mucosa

The inner layer of the alimentary canal is the **mucosal layer**, and is composed of the luminal epithelium, which will come in contact with the processed food. A variety of cells can be found in this layer, depending on the function of the particular region. Beneath the epithelium, you will find the **lamina propria**, which has loose connective tissue and an abundance of lymphocytes. The boundary of the mucosa is a thin layer of smooth muscle called the **muscularis mucosa**, which functions in the mechanical processing of material.

### Submucosa

This region of connective tissue underlies the mucosa and is where you will find blood vessels, lymphatics, and nervous system **plexuses** that control the muscular contractions (**peristalsis**) of the canal. Meissner's (**submucosal**) plexus is located in this layer and provides parasympathetic control of the various secretions in a particular region. This is part of the autonomic nervous system, particularly the enteric nervous system that controls the alimentary canal.

### Muscularis

Made up of thick layers of smooth muscle, this third layer facilitates the peristaltic movement of each region. While variations may occur, the general structure of the muscularis is an inner layer of smooth muscle and an outer layer of longitudinal muscle. When the circular muscle is contracted, material in the alimentary canal will be segmented in much the same way that a baker kneads dough. The material may be pinched in half or otherwise compressed in both directions as the canal narrows in that region. Along with the constriction of the circular muscle, the longitudinal muscle will act to pull portions of the canal closer to each other. Rhythmic contractions of both circular and longitudinal muscle layers will move material progressively down the alimentary canal.

These contractions are controlled by another component of the enteric nervous system, the **Auerbach's (myenteric) plexus**. This is located between the circular and longitudinal smooth muscle layers.

### Serosa/Adventitia

The outermost connective tissue of the alimentary canal is the **adventitia**. This connective tissue allows the alimentary canal to be secured to the connective tissue of the body wall in certain areas. In other locations, the canal is not attached to the body, but is covered by a thin layer of **mesothelium** called the **visceral peritoneum**. In this case, the outer layer is called the **serosa**.

A mesothelium is a thin layer of flattened cells that provides a covering and protective layer for internal surfaces of the body. The mesothelium on the surface of internal organs is called the visceral peritoneum, while the mesothelium that lines the thoracic and abdominal cavity walls is called the parietal peritoneum.

## Pharynx and Esophagus

As food becomes mixed with saliva, it is processed by the teeth and tongue and turned into a spherical mass called a **bolus**. This bolus is moved to the back of the mouth in preparation for swallowing. From the oral cavity, the bolus is moved into the **oropharynx**, which is commonly called the throat.

If the pharynx were considered a pipe, then in plumbing it would be a T intersecting pipe. The oropharynx would be the stem of the T and connects to the upper portion descending from the nasal cavity (**nasopharynx**) and the lower portion connecting with the larynx and esophagus (**laryngopharynx**).

As the food is swallowed (deglutition), a cartilaginous flap, the **epiglottis**, reflexively covers the **glottis** (opening of the trachea) and prevents material from being aspirated into the airway. Additionally, the epiglottis creates a ramp to help direct the bolus into the esophagus.

Composed of an epithelium that is identical to that of the oral cavity, the esophagus will utilize its muscularis to propel food downward and into the stomach. The upper portion of the esophagus contains a high proportion of skeletal muscle in the muscularis layer, and this can be under either voluntary or involuntary control. However, there is a gradual and consistent transition from skeletal to smooth muscle as the esophagus descends. By the time the esophagus connects to the stomach, the muscle is 100 percent smooth muscle, a portion of which will become the lower **esophageal sphincter**, which prevents regurgitation of stomach contents, including acid, upward into the esophagus. The esophagus passes through an opening in the **diaphragm** (hiatus) to connect with the stomach in the upper abdominal cavity.

The muscle composition of the esophagus differs as it descends to the stomach, and so does its function. For instance, with 100 percent skeletal muscle at the upper portion of the esophagus, the swallow can be voluntarily controlled. However, after the swallow is initiated and as the increasing concentration of smooth muscle takes over as the material descends, the movement of the material becomes governed by involuntary control and can't be stopped.

## Stomach

Capable of expanding from a few cups when empty to more than a gallon when completely filled (you will feel full at around 4 cups), the stomach is the location where mechanical and chemical digestion is increased. Examination of the interior lining of the stomach will reveal that folds called **gastric rugae** occur when the stomach is empty. These folds accommodate the expansion of the organ. Interestingly, the use of hydrochloric acid by the stomach presents a unique problem: how to digest food without digesting your own tissues. A variety of cells in the stomach accomplish all of the tasks required to do this.

## Cells

At the cardia of the stomach (the location of attachment of the esophagus to the stomach), the mucosal lining cells change from the epithelium of the esophagus, which resists friction, to an epithelium that will line the stomach and resist the harsh chemical environment. Therefore, **surface lining cells (SLCs)** line the luminal surface of the stomach. These cells have tight junctional complexes between adjacent cells, creating a watertight barrier to keep stomach contents from leaking into the underlying tissue. If you were to examine the stomach epithelium, you would notice depressions of the mucosa throughout the organ. These gastric pits are invaginations of the epithelium downward through the **lamina propria**. They populate the entire mucosal layer. This vastly increases the surface area of the epithelium and creates a protected environment in the pits for the secretory cells.

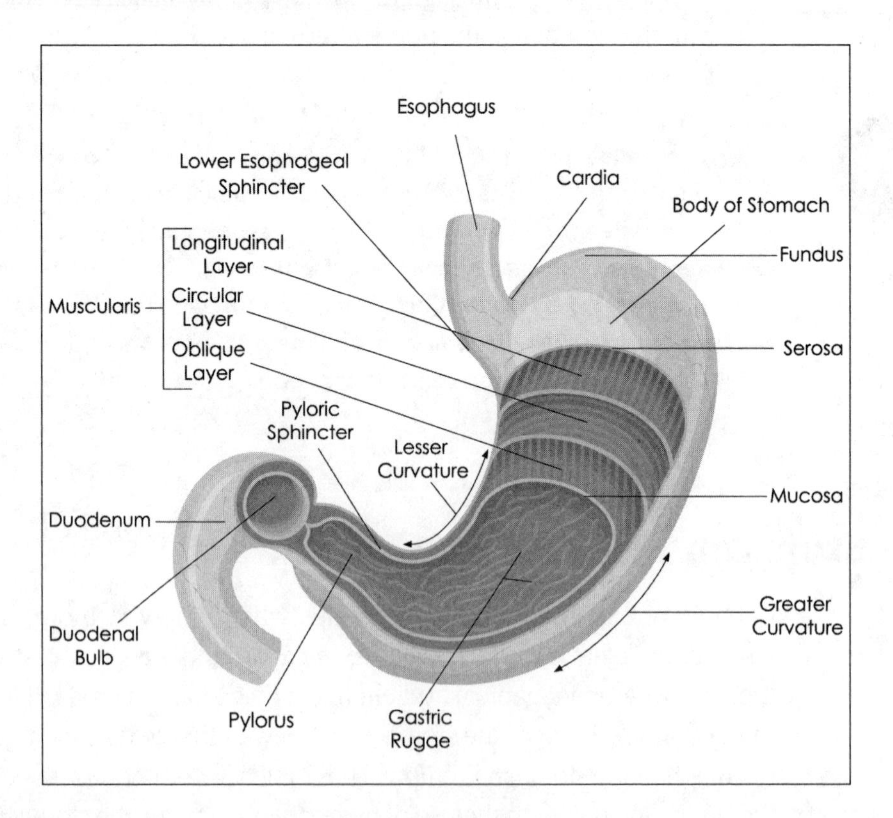


Notice the internal folds (rugae) of the stomach that allow the stomach to expand greatly during eating and drinking.

The first portion of the gastric pits is referred to as the **neck**. The cells in this area are mucus neck cells that produce a visible mucus, rich in bicarbonate, which coats the SLCs and protects them from the harmful HCl in the stomach. As you descend farther into these pits, **parietal cells** become more numerous. These are the HCl-producing (oxyntic) cells of the stomach. Also populating the base of the pits are **chief cells**, which produce and secrete an enzyme-rich mixture and are classified as **zymogenic cells**. The final category of cells found in the stomach is the **enteroendocrine cells**, which produce a number of hormones in response to stomach activity including glucagon (to mobilize liver glycogen), gastrin (to signal HCl production), and serotonin (to stimulate stomach peristalsis).

### Divisions

In the cardia of the stomach, gastric glands that are rich in mucus-producing cells to protect against the harmful acid in the stomach predominate. However, in the **fundus** of the stomach, the dome-shaped superior portion, and the body of the stomach, the gastric glands (**fundic glands**) are densely populated with parietal cells. The most inferior region, just before joining the small intestine, is the **pylorus**, which, like the cardiac region, is rich in mucus-producing cells that will neutralize the stomach acid before transferring stomach contents to the intestine. Finally, at the junction of the stomach and intestine, the inner circular layer of the stomach is expanded into the **pyloric sphincter**, which will regulate the passage of material from the stomach into the first portion of the small intestine (duodenum).

**What is occurring when your stomach "growls"?**
The sounds heard and vibrations felt when the stomach growls happen because of the muscular contractions of the stomach when it is empty and is moving air between the folds (a.k.a. rugae) of the stomach.

## Small Intestine

The small intestine is the longest portion of the alimentary canal, reaching approximately 20 feet in length. While the stomach facilitates mechanical and chemical digestion, the small intestine is largely tasked with absorbing the raw digested materials into the blood and lymphatic system.

To do so, the luminal mucosa has been expanded and folded to increase surface area. The mucosa is expanded out into the lumen as folds or finlike protrusions (**plicae circulares**) that look and function much like the blades on the inside of a clothes dryer (which helps move material around). Furthermore, smaller folds are present over the entire mucosal surface called **intestinal villi** (see figure of small intestine). This is where absorptive cells are located. Additionally, apical modifications called **microvilli** on the lining cells maximize the absorptive area. These 3 mechanisms together increase the area 600-fold.

In addition to increasing surface area, the intestinal villi also provide a space within the lamina propria of each villi for vascular capillaries and lymphatic capillaries. Since the lining cells will pass material from the lumen to the internal environment of the villi, these vessels will quickly remove the material from the intestinal tract and pass it on to the body.

Between the intestinal villi and extending downward into the lamina propria are extensions of the mucosal epithelium. These may be considered similar to the gastric pits in the stomach; however, these are basically the tight spaces between the surrounding rising intestinal villi and are therefore named the **crypts of Lieberkühn** (intestinal crypts). Here you will find cells with functions other than absorption, including the secretion of hormones, enzymes, and acid-neutralizing mucus.

## Cells

The predominant cell type found lining the intestinal tract is the **surface absorptive cell (SAC)**. With numerous microvilli, these cells are well suited to absorb material from the intestinal lumen. Additionally, with tight junctional complexes between cells, they ensure that no material in the lumen will pass into the deeper tissues of the intestine.

As material is passed along the alimentary canal, it will become progressively less hydrated and, as a result, more difficult to move without causing damage to the tissues. Therefore, mucus-secreting cells (goblet cells) will become progressively more abundant as a component of the mucosal lining.

As found in the gastric pits, the crypts will also be populated with enteroendocrine cells that help produce hormones. Additionally, these cells will produce **gastric inhibitory peptide** (to stop the production of HCl) and **cholecystokinin** (causing peristaltic contractions of the gall bladder to expel bile into the alimentary canal).

In the **ileum** of the small intestine, lymphoid nodules are observed in the lamina propria. Just above these permanent nodules are **M (microfold) cells** in the epithelium. They are antigen-presenting cells, much like macrophages, and pass luminal material directly to the underlying nodules in the ileum.

The most histologically characteristic cells of the intestinal tract are the large **Paneth cells** present in the base of the intestinal crypts. It is thought that encounters with pathogens trigger these cells to release their secretory materials. Paneth cells produce a variety of antimicrobial enzymes and agents such as **lysozyme** and release immune system cytokines that are essential for immune system function.

## Divisions

The small intestine is divided into 3 regions. The first, and the shortest, is the **duodenum**, which is connected to the pylorus of the stomach. Receiving material directly from the stomach, this region has an abundance of cells that produce an acid-neutralizing mucus. Secretions from **Brunner's glands**, which are present in the submucosa, also accomplish this. These glands transport their acid-neutralizing mucus to the lumen. Therefore, Brunner's glands, found only in the short duodenum, are the histological feature aiding in the identification of this tissue. Additionally, material from the pancreas and bile from the liver are introduced into the duodenum from the common bile duct to finish chemical digestion in a more neutral environment.

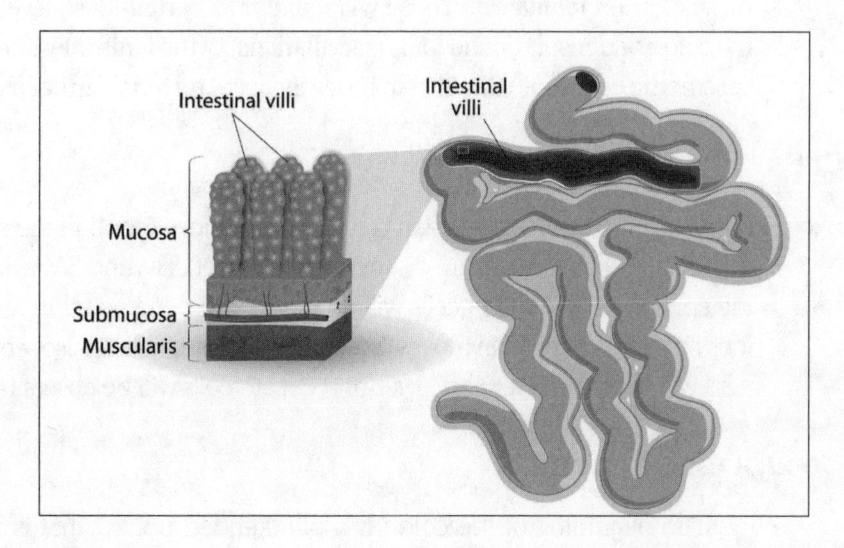


The small intestine is highly coiled. The internal surface area of the lining epithelium is increased by numerous fingerlike protrusions called intestinal villi, which project into the lumen of the small intestine.

The **jejunum** is the middle region of the small intestine and is approximately equal in length to the last portion of the small intestine, the ileum. **Peyer's patches** are the histological fingerprint for the ileum. These are the permanent lymphoid nodules found in the lamina propria of the ileum.

## Large Intestine

The large intestine (colon) functions primarily in absorption, absorbing water (approximately 1400 ml/day). As the material passing through the large intestine becomes more and more dehydrated, it is compacted into solid waste (feces; 100 ml/day) and is stored in the lower portion of the colon until eliminated from the body. Digested material passes from the ileum of the small intestine through a muscular sphincter (**ileocolic valve**) and into the beginning of the colon.

The first big difference observed between the colon and the small intestine is the absence of intestinal villi. The crypts will still be present, but will become shorter as the fecal material gets closer to the rectum (end of the large intestine). While the colon is not the longest portion of the alimentary canal, it still stretches for approximately 5 feet and is larger in diameter than the small intestine (3 inches versus 1-inch diameter of the small intestine).

Another important histological, as well as gross anatomical, difference is that the outer longitudinal muscle in the colon is only present as 3 bands of smooth muscle called **taenia coli**. These will maintain a certain base level of tension on the colon and results in the folds (**sacculations**) of the large intestine. Additionally, spasmodic contractions will help move the fecal material farther along the colon.

### Cells

While the cells in the colon are not unlike those found in the small intestine, their relative proportions are distinct. While goblet cells and SACs are both part of the epithelium, the goblet cells will dominate and become more and more populous as the fecal material moves closer to being eliminated. Also, enteroendocrine cells will become fewer and fewer, and Paneth cells will be absent from the colon.

### Cecum and Appendix

The beginning of the colon is a blind-ended pocket that is inferior to the ileocolic valve called the **cecum**. Material will be moved upward into the first ascending portion of the colon; however, some material will invariably be trapped in the cecum, which is not unlike a dead-end street. Therefore, projecting off the cecum is a worm-like appendage called the **appendix (vermiform appendix)** where lymphoid nodules can be found in the lamina propria.

### Colon

From the cecum, the **ascending colon** rises superiorly on the right side of the abdomen before making a 90° bend and extending across the body as the transverse (across) colon. Once on the left side of the body, the colon makes another 90° turn and forms the descending colon, which extends to the lower left quadrant of the abdominal cavity. To align with the midline of the body, the colon makes an S-shaped bend as the sigmoid colon and then continues straight downward as the rectum.

### Rectum

The final straight segment of the colon is the rectum, which functions in the storage of feces and its elimination (defecation). Just before the **anus** (external opening of the rectum), 2 muscular sphincters retain the material internally until voluntarily released. The internal anal sphincter is under involuntary control and is always in a state of contraction. It is triggered by fecal pressure. The external sphincter may be controlled voluntarily or involuntarily, and is made up of skeletal muscle.

## Pancreas

Considered an accessory digestive gland, the pancreas is located in the curve of the duodenum near the pylorus of the stomach. This is a perfect location, since the exocrine secretions from the pancreas will pass through the pancreatic duct into the common bile duct, then enter the duodenum through the **sphincter of Oddi**.

During fetal development, the intestines grow faster than the abdominal cavity and will actually herniate (protrude) into the umbilical cord until the body grows large enough to accommodate their mass. At this point, they are packed back inside the body. .

### Exocrine Portion

Triggered by hormones like **cholecystokinin** (produced by enteroendocrine cells of the intestine) and the neurotransmitter **acetylcholine** (active during the rest and digest phase of the autonomic nervous system), the pancreatic cells will secrete a solution of digestive enzymes as pancreatic juice. As much as 1200 ml of this fluid may be produced daily; it contains enzymes that further degrade carbohydrates, proteins, and fat.

### Endocrine Portion

These cells are the characteristic feature of the pancreas. Aggregated into small masses, these endocrine cells form the **islet of Langerhans** (pancreatic islets) and are surrounded by the exocrine cells of the pancreas.

## Liver

The liver, the largest gland in the body, is positioned in the superior portion of the abdominal cavity just superior to the stomach. It is divided into 2 major lobes (right and left) and 2 minor lobes (the **quadrate lobe**, located near the gall bladder, and the **caudate lobe**, near the entry of the hepatic portal vein). Blood from the digestive tract is brought to the liver via the hepatic portal vein and enters the liver at the junction of the 4 hepatic lobes, called the **porta hepatis**. This nutrient-rich blood is spread throughout the open spaces of the liver (sinusoids) and forced to make contact with the liver cells (a.k.a. hepatocytes), which can metabolize this material.

### Structure

When considering the structure of the liver, one can think of it as a high-efficiency filter. Blood is emptied into long channels between rows of liver cells (hepatic cords) and flows toward a central vein. With these rows of cells and sinusoids arranged in series, it creates what resembles a 6-sided wheel with the spokes (rows of cells) radiating toward the axle (central vein). At these peripheral points on this hexagon, there will be an arrangement of a hepatic vein (a branch of the hepatic portal vein), a hepatic artery, and a bile duct. These will always occur together at these points and are thus referred to as a **hepatic**

**triad**. Thus, blood may flow from the hepatic artery (oxygen-rich blood) and the hepatic vein (oxygen-depleted but material-rich blood) through the sinusoidal spaces inward and finally be drained by the central vein.

As this mixed blood travels through the hepatic sinusoids, it is in contact with the capillaries that line the spaces and separate the blood cells from the underlying liver cells. However, the endothelial cells of these capillaries are filled with large holes and resemble Swiss cheese where there are more holes than cells. Their location and porous appearance have resulted in them being named sinusoidal capillaries. These large holes allow all material in the blood, with the exception of cells and platelets, to pour into a space (**space of Disse**) between the sinusoidal capillaries and the hepatocytes. It is within this space that microvilli from the hepatocytes project upward to maximize the surface area in contact with the leaked plasma for absorption and metabolism of the materials.

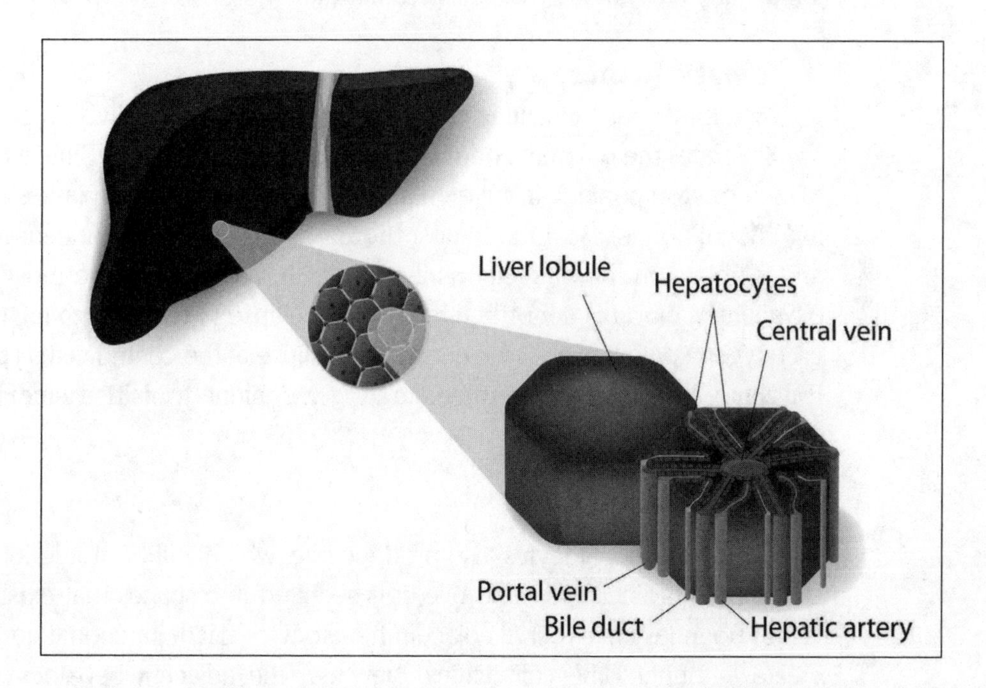


The functional division of liver cells (hepatoctyes) into a hexagonal classic lobule.

As you might imagine, material coming from the digestive tract may contain less desirable material. Thus, resident macrophages, called Kupffer cells, are abundant in the liver sinusoids and remove any substances that may be detrimental to the body.

## Lobules

With a foundational knowledge of the cellular and tissue organization of the liver, the functions facilitated by the liver may be mapped to the tissue into one of three conceptual divisions called lobules.

### Classic Lobule

The basic structural and functional unit of the liver is the classic lobule. The hexagon-shaped wheel functions to move blood through the sinusoids, where it can be metabolized by **hepatocytes** (i.e., liver cells). From there it moves into the central vein, where it will eventually be removed by the **hepatic vein**, and then into the **inferior vena cava**.

The following 2 lobules are conceptual in nature and are visualized as different templates laid over the classical lobule.

### Hepatic Acinus

Also known as the **acinus of Rappaport**, this functional division of liver tissue illustrates the different zones of blood flowing through the sinuses relative to their oxygen content. It is drawn as an ellipse (a curve on a plane surrounding 2 focal points), with the 2 most distant points as 2 adjacent central veins, and 2 points on the diameter as 2 adjacent **hepatic triads**. The most heavily oxygenated blood is near the middle of the ellipse (**periportal zone**), the lowest oxygen content is near the ends of the ellipse at the central veins (**pericentral zone**), and a zone of intermediate oxygen content (**transition zone**) is in the middle.

### Portal Lobule

Finally, the hepatocytes also produce bile, which will be stored in the gall bladder. As the cells produce bile, it is secreted into spaces that exist on the lateral boundary between 2 adjacent hepatocytes. Tight junctional complexes create a tunnel (**bile canaliculus**) between the adjacent hepatocyte membranes, which is used to transport bile to the periphery of the portal lobule and into a bile duct. This functional lobule is visualized as a triangle where the points are 3 central veins with a hepatic triad in the center. While the 2 other zones illustrate blood flowing from the periphery to the central vein, the portal lobule illustrates bile flowing through bile canaliculi from the hepatocytes toward the bile duct (from center to periphery of a classical lobule).

## Functions

In addition to blood filtering and bile producing activities, the liver is also a storage facility for carbohydrate (glycogen) and vitamins (A, D, and B). This requires that the liver be able to process carbohydrates. It can polymerize glucose into glycogen, and conversely break down the glycogen it stores and return glucose to the blood stream.

The liver also plays an important role in protein metabolism and composition of plasma. First, the liver produces albumin, which is the most abundant plasma protein. This protein is critical for the maintenance of osmotic pressure in the blood stream. Additionally, a byproduct of red blood cell destruction and hemoglobin metabolism is bilirubin. Bilirubin is further processed by the liver into a water-soluble form that can be eliminated through the urine and feces.

**What is jaundice?**
Jaundice is the buildup of bilirubin in the tissues of the body, resulting in a yellowish tint to the skin and eyes. Bilirubin builds up in the tissues because of lack of liver function, either from disease or from premature birth.

Fats are also metabolized and regulated by the liver, and cholesterol and lipoproteins are produced by the liver. High-density lipoprotein (HDL) and low-density lipoprotein (LDL) are often considered the "good" and "bad" cholesterol, respectively, for their function in either removing or depositing cholesterol into the blood vessels of the body. Your cholesterol content may not be related so much to your diet or level of activity as to your genetics. Your inherited genes dictate the liver enzymes and the level of molecules produced that function in cholesterol deposition and metabolism. Too many or too few of these essential metabolic regulators can lead to unhealthy blood levels of cholesterol.

Lastly, the liver is best known for its ability to detoxify the blood. Like a sponge, the liver removes toxins from the blood and enzymatically processes the materials into a less harmful form. Ethyl alcohol is the best example of the material detoxed by the liver. If you consume alcohol at a rate that exceeds what the liver can metabolize, your blood alcohol levels increase and coordination and judgment decrease. Chronic alcohol use is the leading cause of **cirrhosis** of the liver, which is a permanent loss of liver tissue and function.

## Nutrition

Although the human body is very good at transforming materials into usable substances and storing excess energy in reserves for later use, it still relies on environmental sources for fuel and materials to maintain homeostasis. For example, muscle is basically proteins organized in such a way as to contract and facilitate work. These materials do not last for a lifetime and are in a continuous state of remodel and repair. Therefore, protein must be obtained from dietary sources to provide this critical building block for human function.

Again, the body can transform existing materials into some amino acids (the unit of protein structure); however, some essential amino acids can only be obtained from digested food. Without a sufficient supply of dietary protein from either meats or vegetables, the human body will atrophy, growth will stop, and eventually premature death will occur. A diet of proteins alone will not provide a full array of essential materials for healthy body function.

Carbohydrates have been much badmouthed in the media and by weight-loss proponents. However, carbohydrates are the foundational fuel for all cellular activity in the body. Without carbohydrates, the body will turn to lipids and protein as survival fuel. In that case, byproducts will accumulate in the blood stream, resulting in a lowering of blood pH (metabolic acidosis). Dizziness, confusion, and muscle weakness are all side effects of acidosis.

That being said, not all carbohydrates are good for your health. Those carbohydrates that are broken down and flood the blood stream rapidly (high glycemic index carbohydrates) cause a spike in blood sugar, which is hormonally reduced because the excess high sugar is stored as glycogen and fat. Additionally, in a short period of time, the blood sugar will rapidly drop (crash) and the individual will feel weak and lethargic as the body now tries to shift metabolic gears and put stored sugar back into the blood stream. The sugar roller coaster is not a healthy way to utilize carbohydrates in the diet. These high glycemic index sugars include processed sugars, white breads, white rice, potatoes, and sugary beverages. Whole wheat bread, fruits such as apples or pears, and vegetables such as carrots are foods with a low glycemic index. These will enter the blood stream more gradually as the materials are absorbed across the intestinal wall and maintain a rather steady state of blood sugar, not causing the sugar spike and storage emergency (and crash).

In addition to eating foods rich in protein and lower in carbohydrates (and containing the healthier carbohydrate choices), the amount of fat your body takes in should be minimized. Notice that the recommendation wasn't to eliminate fat completely. Dietary sources of fat and cholesterol are important in the production of new cell membrane and steroid hormones for the body. Certainly excess fat consumption can lead to many health issues, including coronary artery disease; so, fat intake must be carefully monitored and regulated.

While the composition of the foods we eat is an important aspect of nutrition and health, the amount of what is eaten when compared to level of physical activity is often overlooked. To maintain an ideal weight, a healthy diet should include no more calories than are demanded by the body on a daily basis. Any amount of calories over this will result in fat stores increasing in the body. An additional 500 calories consumed daily (that are not used for energy) will result in the addition of 1 pound of fat over the course of a single week. Likewise, a daily reduction of 500 calories per day will reduce a pound of fat in a single week. Therefore, diets rich in protein, with carbs of low glycemic index, and limited in fats, along with careful control of the amount eaten compared to energy used, will yield a healthy lifestyle that can be maintained for a lifetime.

## Diseases and Disorders

With such an extensive and diverse system as the digestive system, the problems, malfunctions, and diseases are numerous. Following are just a few of the digestive problems that you may encounter.

### GERD

A growing disorder in Westernized cultures is **gastroesophageal reflux disease (GERD)**. Basically, this is a chronic heartburn condition. Gastric acid, normally restricted to the stomach where an acid-neutralizing mucus protects the surface lining cells from damage, regurgitates past the lower esophageal sphincter and into the esophagus. Cells of the esophagus are not protected from the acid and the result is a burning sensation. Long-term exposure of the esophageal tissue to acid may lead to esophageal cancer.

GERD is often a result of an expanded hiatus or even a hiatal hernia, where portions of the stomach project through the hiatus in the diaphragm and into the thoracic cavity. Normally, the lower esophageal sphincter is at the same level as the hiatus, and as a result the muscle of the diaphragm provides extra support for closing this important valve and keeping acid in the stomach. Repair of such a defect is often enough to resolve the GERD.

In other cases, an overproduction of acid, diet, or lifestyle can increase the occurrence of acid flowing into the esophagus. Medications including proton pump inhibitors may be prescribed to limit the stomach's production of HCl. Additionally, eating smaller, more frequent meals and not lying horizontally immediately after a meal are lifestyle alterations that can minimize the bouts of GERD.

## Peptic Ulcers

While the stomach has natural means of protecting its own cells from the harmful effects of acid, in some cases damage occurs to the mucosal wall and allows the acid to penetrate into the underlying connective tissue and cause further damage. Sensitive pain receptors present in the submucosa alert the individual of pain, especially after eating a meal when the acid increases. In many cases, this wound fails to properly heal due to populations of indigenous bacteria (H. pylori) that collect in the wound site. Antibiotics are often effective at resolving these minor ulcerations and allow the body's wound healing process to close the mucosa.

## Diarrhea

While this condition may just seem like a minor inconvenience to some, for many throughout the world it is a life-threatening condition. According to the World Health Organization (WHO), over three-quarters of a million children under the age of 5 die from diarrhea each year. The intestinal crypt cells typically produce a secretion daily that contains antibacterial enzymes (intestinal juice). When pathogens or parasites are detected in the intestine or mucosal layer, these glands shift into overdrive and produce massive amounts of fluid to flush the alimentary canal. Additionally, absorption is reduced to allow more material to flush the system in hopes of removing the problematic materials. If the diarrhea is not alleviated and if fluids cannot be retained, dehydration will occur in a short period of time and may eventually lead to death if severe enough.

## Hepatitis

Literally meaning inflammation of the liver, hepatitis has many causes, including sexually transmitted viruses (the most common cause), chronic alcohol consumption, and autoimmune diseases. During the inflammatory process, immune system cells flood the liver tissue and essentially interfere with normal liver function. Therefore, the symptoms resulting from liver damage include yellowing of the skin (jaundice), nausea, vomiting or diarrhea, and loss of appetite.

Depending on the cause of the disease, the prognosis is varied. With chronic damage to the liver, scar tissue will build up and block the regenerative ability of the liver and result in permanent damage and loss of function. Therefore, prevention is highly advised (this can be accomplished with a vaccine). Early administration of vaccines for hepatitis A and B yielded great success (90–100 percent efficacy) in preventing the contraction of this disease.

CHAPTER 15

# Respiratory System

The role of the respiratory system is rather simple: bring oxygen ($O_2$) to the tissues and remove carbon dioxide ($CO_2$) from the body. This system is divided into the conduction zone, where the air is transported but gases are not exchanged with the blood, and the respiratory zone, where gases are passed between the airways and the blood. Additionally, you'll learn about the anatomical arrangement of the respiratory system, the muscles actively involved in ventilation (breathing), the movement of gases by diffusion into or out of the blood, and finally, the effect of these gases on blood and tissue pH.

## Respiratory Epithelium

The lining of the respiratory tract is meant to warm, humidify, and clean the air as it passes down the tract before entering the lungs. To do so, there are specialized cells that facilitate these functions. Remember that epithelium is layers of cells that line hollow organs and glands, and make up the outer surface of the body. In the case of the respiratory tract, the epithelium is arranged into what most consider the typical **respiratory mucosa**, which is a collection of cells that give the appearance of being multilayered. Every cell will have an attachment to the basement membrane; every cell does not, however, reach the lumen, and this results in nuclei being at different levels. This is a **pseudostratified columnar epithelium**.

### Columnar Epithelium

The surface epithelial cells are taller than they are wide so they are considered to be columnar in appearance. This shape is simply due to their arrangement in the layer and does not have a practical significance otherwise. However, modifications to their **apical** (facing the lumen or surface) membranes play a crucial role for the respiratory system.

Extensions of the cell membrane are projected into the lumen as cilia. These slender processes have a core of **microtubules** (cytoskeletal fibers) that are arranged in such a way as to be anchored to the cytoskeleton at the base (**basal body**) and freely extend outward as the main body or shaft of the cilia.

Cilia are often compared to hairs, but are completely different in molecular organization and scale (hairs are made up of cells and cilia occur on single cells in the hundreds).

Using the **microtubule-associated proteins** (MAPs) dynein and nexin, the cores of the microtubules can be made to slide relative to each other, resulting in bending of the cilia. When the dynein completes the bending cycle, which requires ATP, the resulting relaxation and rebound is created by nexin, which is rather elastic in nature. This active bending and rebounding causes the cilia to wave back and forth and creates a current that is capable of moving surface material. When working in concert, all the cilia on all the columnar cells of the respiratory tract can beat to move material up and out of the respiratory tract. A sticky substance called mucus that can trap the particulate material aids this movement to more efficiently transport the material.

### Goblet Cells

These are the same cells as were mentioned in the digestive tract, and they produce a thick, sticky mucus just as is done in the alimentary canal. However, while their function in the digestive tract is lubrication, in the respiratory system the mucus acts as a dust trap to grab particles from the air so they can be more efficiently transported upward and eventually out of the respiratory tract.

## Nose

The nasal cavity is one of the two ways air can enter the body, the other being the mouth. Air passes through the nostrils (external **nares**) and remains either on the right or left side due to the nasal septum. Once in the cavities themselves, the air first encounters the respiratory epithelium, the first stage in the process of being cleared of particulates. Additionally, in this pathway especially, the air is warmed and humidified. To ensure maximum efficiency in all of these processes, the walls of the nasal cavity have several large folds (nasal conchae). Boney plates create the internal mucous folds that project outward into the nasal cavity and expand the surface area where air can be processed before moving farther along the respiratory pathway.

The autonomic nervous system creates a nasal cycle where the majority of the air passes through one nasal cavity at a time on an alternating basis. You notice this more when one side of your nose is congested.

The air then moves toward the back of the nasal cavity and through narrow passages (internal nares) that lead to the top portion of the throat (**pharynx**). This **nasopharynx** connects the nasal cavities to the **oropharynx**, which is the back of the oral cavity. This is where you would find your pharyngeal tonsils (adenoids), collections of immune system cells that are often removed at the time when the tonsils (palatine tonsils) are removed from the oropharynx.

## Pharynx and Larynx

Commonly referred to as the throat and voice box, the pharynx and larynx respectively mark the beginning of the respiratory tract and prevent the aspiration of liquid and solid material into the airways. In addition, you will find the structures that are essential for vocalization (sound) in this area.

Air from the nasopharynx and/or the oropharynx must make its way downward to the **laryngopharynx**. This cartilaginous chamber is protected from liquid and particulate matter by the overlying cartilaginous flap, the **epiglottis**. While liquid or food causes the epiglottis to fold over the opening of the larynx (glottis), air easily finds its way around the epiglottis and moves into the larynx. Within the larynx, connective tissue shelves on the lateral portions of the passageway and muscles will create the adjustable cords that vibrate as air moves between them, creating sound. The most superior folds (shelves or cords) are thicker and more substantial in structure to provide a protective umbrella for the more delicate cords that are inferior. These superior folds are called the **vestibular folds** and they protect the vocal folds (vocal cords).

It is these delicate, thin connective tissue shelves that stretch across the laryngeal opening and become taut or loose to provide a larger or smaller opening, which changes the pitch of the sound emanating from the vibrating cords. The **vocalis muscle**, which is lateral to the folds in this region, directs this tension on the cords.

## Trachea and Bronchial Tree

The trachea (windpipe) marks the single passage for all air to get to and from the lungs. The trachea lies just under the esophagus and can be viewed on the surface of the neck. This single pipe will branch off a number of times, some outside and most inside of the lung tissue itself. With each branch, the passageway will become considerably narrower and the walls of the tubes will become thinner until the dead end is reached and gases can be exchanged between the air and the blood.

The trachea is formed with intermittent cartilaginous rings that occur along the length of this organ. These incomplete, C-shaped rings will provide support for the trachea, especially when you inhale, where the lower pressure would likely cause a less well-supported tube to collapse inward on itself. Additionally, at the ends of the "C" there is a muscle that can pull on each end and make the ring (and the diameter of the trachea in that area) smaller. This muscle is called the **trachealis muscle** and can be found at this portion of each of the 16–20 tracheal cartilages.

The first branching of the **bronchiolar tree** (the branched air passages) is outside of the lung tissue (extrapulmonary). This splitting up results in a right and left primary **bronchus** (bronchi is plural). As you will see in the next section, the right lung has 3 lobes and the left has 2 lobes. Therefore, to provide greater airflow to the side with more lung tissue, the right bronchus is larger in diameter than the left.

Any aspirated object is more likely to lodge in the larger-diameter right bronchi than the left.

Primary bronchi will divide and give rise to secondary bronchi within the lung tissue dependent upon the number of lobes of the lungs on each side. This division happens within the lung tissue itself and marks the first **intrapulmonary bronchi**. The right bronchus divides into 3 secondary bronchi, each going to each lobe in the right lung, and the left bifurcates into 2 secondary bronchi for each of the lobes of the left lung.

**Tertiary bronchi** are the next branches that occur on the tree and provide air to a discrete and isolated section of the lung tissue called a **broncopulmonary segment**. Connective tissue septa separate these regions from other segments of the lungs. They are completely independent, with their own vascular supply and airway. On the right side, there are 10 segments of the lungs, and 8 segments on the left with each lobe divided into 4 segments each.

These 3 levels of bronchi will end in the next order of branching where the passageways become **bronchioles**. Bronchioles are much smaller in diameter than bronchi and lack cartilaginous rings or plates, which is a key histological feature of the bronchi. However, layers of smooth muscle will wrap around the ever-narrowing bronchioles to enable them to constrict or dilate as dictated by the autonomic nervous system. From the **tertiary bronchi**, there will be approximately 17 orders of branching continuing until the end of the conduction zone is reached and the respiratory zone begins.

The **terminal bronchiole** marks the last segment of the conducting zone (hence the name terminal). The diameters of these passageways are very thin with only one layer or two of smooth muscle still present. However, the structures required for gas exchange are not present at this level; therefore, no respiration will occur. This will occur in the deepest regions of the lungs.

# Lungs

Divided into right and left lungs, the tissue that makes up these essential organs is more empty space than actual structure. Take away the bronchiolar tree, blood vessels, and nerves and what remains will very much resemble an extremely porous sponge. This thin-walled and porous area marks the bulk of the respiratory zone, where gases can be exchanged between the air and the blood.

## Respiratory Zone

From the terminal bronchioles, air passes to the first portion of the respiratory zone, which is called the **respiratory bronchioles**. These are often difficult to identify histologically because their structure is the same as the terminal bronchiole with one critical exception: the presence of **alveoli**. These are millions of small bubbles that mark the end of the respiratory zone for the lungs. Composed of flattened epithelial cells and having a diameter of approximately 0.5 micrometers, these structures mark the area of gas exchange.

There are 300–500 million alveoli in the lung. Scientists estimate that these millions of alveoli make up approximately 30–50 meters of surface area for gas exchange. This is about the size of a standard tennis court.

Following the respiratory bronchioles, the air moves into an area termed an **alveolar duct**. These thin-walled passages are more defined by the alveoli themselves. However, the openings of the alveoli that protrude off this duct possess a single smooth muscle cell, which creates a cellular sphincter to regulate the air moving into these areas. These are viewed as knobs on the lateral aspect of the alveolar duct.

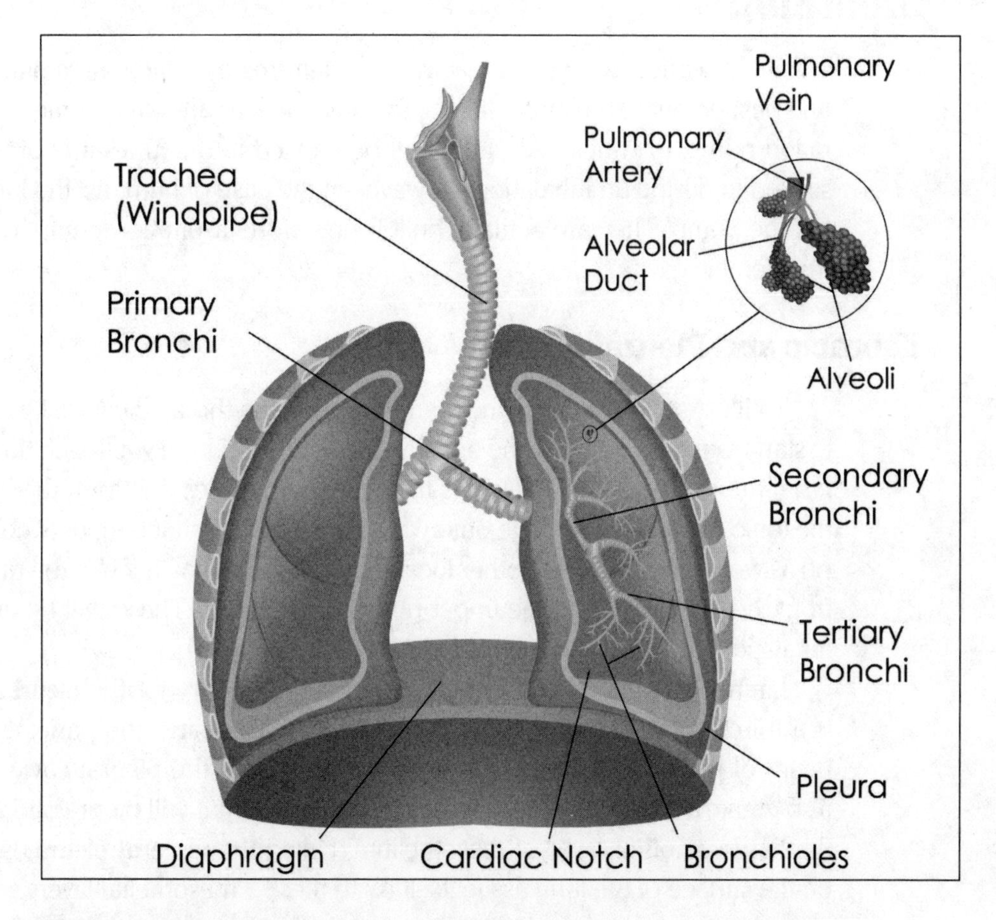


The lower respiratory system is composed of the primary airway (trachea), which conducts air to the lungs and into ever-decreasing diameter passages (primary, secondary, and tertiary bronchi), until the air reaches the small terminal portions called alveoli, where gases are exchanged with the blood.

Finally, air moves into a cluster of alveoli called the **alveolar sac**, which resembles a cluster of grapes. From the common atrium of the sac, air moves into any of the attached alveoli so that gas can be exchanged. Interestingly, there is more than one way to get into or out of an alveolus that is part of an alveolar sac. Between adjacent alveoli, a small opening or pore exists to help maintain the air pressure throughout all the alveoli that are part of the alveolar sac. These are identified as the **alveolar pores** or pores of Kohn.

# Breathing

This next section explains the active mechanisms by which air is pulled into and pushed out of the lungs. In fact, the mechanisms are often termed an aspiration pump, in which a vacuum will be created in the lungs and air will be sucked in during an inhalation. However, in the case of humans, the lungs are not the pump. They are actually built inside of the aspiration pump, which is the thoracic cavity.

## Thoracic and Pleural Cavity

To illustrate this arrangement and describe the thoracic cavity, it is easiest to start from the outside of the body and work inward until you reach the lungs. From the outside, the first layer is the skin and muscles that form the body of the torso. Next is the ribcage onto which muscles are attached, between each rib as well as coming from other locations in the body to attach to the ribs (e.g., from the neck region to the upper ribs of the ribcage). These will be used for inhalation and exhalation (detailed later in this chapter).

Lining the ribcage and forming the outer boundary for the **pleural cavity** is a thin layer of mesothelium called the **parietal pleura** (the parietal peritoneum of the pleural cavity). Not only does this seal the pleural cavity, it will also create a low-friction and nonadhesive surface that will be pressed against the lungs. Another layer of mesothelium, called the **visceral pleura**, is found on the surface of the lung tissue itself. With these 2 mesothelial layers adjacent to each other, intrapleural space is created. As you will see during inhalation and exhalation, the air pressure in the intrapleural space will always be slightly lower than the pressure inside the lungs (intrapulmonary pressure). Therefore, the visceral pleura will always be pressed against the parietal pleura and obliterate any space between the two. This makes it a "potential space" that will only exist if either pleura were to become damaged and the pressure equalize.

**Pneumothorax** is a condition in which damage to the body wall (and parietal pleura) has allowed the intrapleural pressure to equalize with the atmospheric pressure and the lungs pull away from the pleural wall (collapse).

The pleural cavity is basically an isolated cavity that surrounds the lungs, which, under normal conditions, will not receive or lose any air. Within this cavity is the interior of the lungs, which can freely exchange air with the external environment through the bronchiolar tree. Therefore, to cause air to move into and out of the lungs, the intrapulmonary pressure must be lower or higher than the atmospheric pressure, respectively.

## Inhalation

Lung tissue is not muscular and therefore cannot dilate or constrict on its own. The ventilation mechanism for the human body is called an aspiration pump, and is found surrounding the lungs. The pump that drives this ventilation is a collection of muscles in the thoracic cavity. By expanding or constricting the thoracic space, the intrapleural pressure will be decreased or increased. For example, when the thoracic cavity expands, the pressure within the intrapleural space will decrease: No more air can come in, so the expanded cavity becomes a larger space with the same amount of air and the pressure decreases.

Boyle's law states that the pressure of a gas is indirectly proportional to its volume. In other words, the same amount of gas in a bigger space will have a lower pressure and vice versa.

Recall that the intrapleural pressure will *always* be lower than the pressure inside of the lungs. Therefore, where the thoracic cavity goes, the lungs go. For example, expand the thoracic cavity, and the lungs will expand (inhale). The same rule applies to the air in the lungs: Air in a larger space will have a lower pressure. Since the lungs are expanded, and there is lower pressure inside the lungs than atmospheric pressure, air will rush into the airway and enter and fill the lungs.

But how do you expand the lungs to create this larger space and a lower pressure for inhalation? First, consider quiet inhalation, which is what humans do when at rest. One of the main muscles involved in ventilation is the **diaphragm**. This dome-shaped muscle is under subconscious control (but can be voluntarily controlled) and forms the inferior boundary of the thoracic and pleural cavities. When relaxed, the superior dome portion of the diaphragm projects upward into the thoracic cavity. Therefore, when contracted, the diaphragm flattens and moves downward, which increases the volume of the thoracic cavity.

Additionally, one of the 2 sets of muscles between the ribs contracts and attempts to pull the ribs closer together. These muscles are the external intercostals, since they are the most lateral of these sets. The ribs, however, are anchored to the vertebral column and the sternum and therefore can't move closer. This force causes the ribcage to hinge upward, and in doing so expands the thoracic cavity laterally. Both of these actions create the lower pressure required to expand the lungs and thus aspirate air into the lungs during an inhalation.

During periods of activity when respiratory rates increase, inhalation is deeper and more rapid and requires additional muscular support. Support comes in the form of various accessory muscles that are attached near the top of the ribcage and include the **sternocleidomastoid**, **parasternal muscles**, and the **scalenes**. The action of these muscles will more rapidly raise the rib cage for a more forceful inhalation.

## Exhalation

Compared to inhalation, exhalation is simple and passive. To reduce the size of the thoracic cavity after an inhalation, all the muscles involved in inhalation simply need to relax and let gravity pull the ribcage back downward. Also, remember the dome shape of the diaphragm when relaxed? After the inhalation, the diaphragm returns to this shape and further constricts the thoracic cavity, generating high pressure inside the lungs, and therefore causing air to be pushed outward. These are the mechanisms for a quiet or restful exhalation.

As in active inhalation, active exhalation requires additional muscular assistance. The internal intercostals are arranged in a different orientation to the externals such that when they contract they assist in the more rapid lowering of the ribcage and the more forceful exhalation.

## Gas Exchange

At this point in the respiration process, gases will move across the respiratory membrane (by diffusion) and interact with blood elements for transportation into or out of the body.

### Blood-Air Barrier

Air and blood will never naturally mix, but they must come within close proximity to each other for diffusion of gases to effectively and efficiently occur. Therefore, each alveolus is infested with capillaries, which are the only blood vessels that will allow the exchange of gases. The lining cells of the alveoli and the endothelial cells of the capillaries compose the thin blood-air barrier.

In the alveoli, these cells are flattened, much like the capillary endothelial cells, and are called **type I pneumocytes**. Another cell type present at the alveolar level is the **type II pneumocytes** (a.k.a. great alveolar cells). These are huge rounded cells that bulge into the lumen of the alveoli. They do not aid in the exchange of gases, but rather produce a substance called surfactant that will assist in keeping the alveoli open.

At 0.5 micrometers in diameter, the surface tension forces of water would be sufficient to collapse the alveoli inward upon themselves. But the phospholipid-rich surfactant will interact with the water molecules, while at the same time their hydrophobic chains will keep other water molecules at a distance. In this way, the pressure required to keep the alveoli open is all but eliminated in the presence of surfactant.

### External Respiration

External respiration almost sounds like an oxymoron because it happens "in" the lungs. External refers to the gases and their location. In this case, the external air has simply been inhaled into the lungs to the depth of the alveoli. However, the air is still just that, atmospheric or "external" air that must exchange gases with the blood.

As mentioned before, the movement of gases is purely by diffusion. Therefore, one must consider the pressures of the gases in the blood versus the pressure in the air to determine which has the greatest pressure. This will determine the direction of diffusion. Within the alveoli, $O_2$ has a partial pressure of approximately 105 mmHg, while that of the blood (just returning in the veins) is at its lowest at 40 mmHg. This will drive the diffusion of $O_2$ from the alveoli and into the blood. Conversely, $CO_2$ pressure in the alveoli is at 40 mmHg, which is lower than that found in the blood (46 mmHg). This will therefore drive $CO_2$ from the blood and into the alveoli, where it can be expelled from the body with the next exhalation.

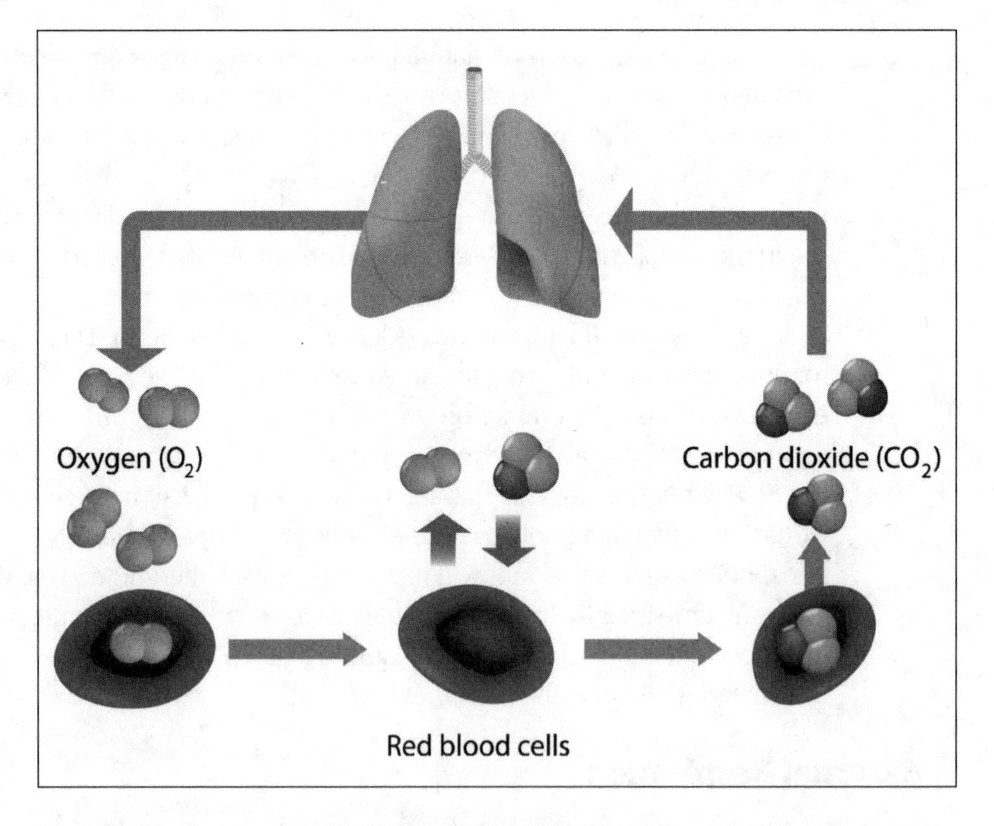


In the lungs, oxygen moves into the red blood cells while carbon dioxide moves from the blood into the lungs in a process called external respiration. This movement of gases is reversed when the blood circulates to the tissues of the body in the process of internal respiration.

## Internal Respiration

Now it is time to transfer the gases to these tissues and cells. Upon leaving the lungs, the pressures of $O_2$ and $CO_2$ are 100 mmHg and 40 mmHg, respectively. At the tissue level, these pressures are lower for $O_2$ and higher for $CO_2$, which drives oxygen out of the blood and into the tissues, and $CO_2$ returns to the blood from the cells. Although the direction is opposite from what occurred in the lungs during external respiration, the direction of movement is still from high pressure to low pressure, thus is moved by diffusion.

# Gas Transport

While some of the gases that dissolve into the blood remain in the liquid plasma, much like the carbonation of a soda is dissolved in the liquid when under pressure, many will be bound and transported (e.g., oxygen to hemoglobin) or may be processed and transported in an alternate form (e.g., $CO_2$ conversion to bicarbonate ion).

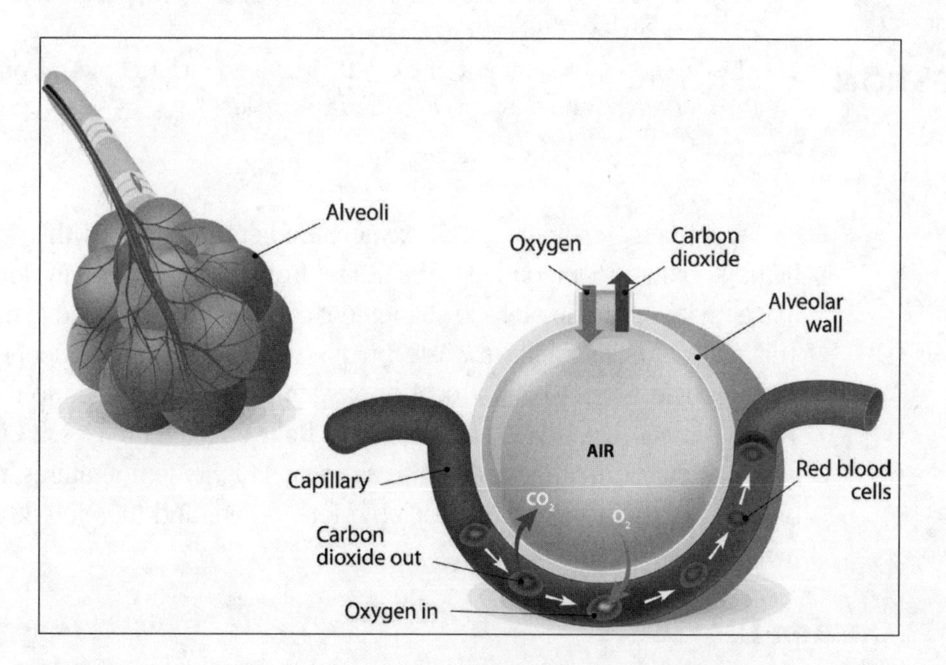


Oxygen is transported in the blood bound to hemoglobin within the red blood cells. Carbon dioxide is processed within the red blood cells, but is converted into carbonate and transported in this form (dissolved in the plasma) until returned to the lungs.

## Oxygen

Red blood cells are the main transportation mode for $O_2$ throughout the human body. Oxygen will diffuse into the red blood cell and will bind to an iron atom, which is held in place by the heme group of the larger protein hemoglobin. Hemoglobin is composed of 4 protein subunits, which in the adult hemoglobin molecule are 2 alpha chains and 2 beta chains (each of which can bind oxygen). Therefore, in the lungs, **deoxyhemoglobin** (hemoglobin lacking oxygen) will bind to oxygen and become **oxyhemoglobin** so it can be transported throughout the body. This is the loading reaction (when oxygen binds to hemoglobin) for hemoglobin, and under atmospheric conditions hemoglobin will exist in a state where 97 percent of the hemoglobin molecules are oxyhemoglobin. In fact, the love hemoglobin has for oxygen is so great that even if the partial pressure of oxygen in the air is decreased from 100 mmHg to 60 mmHg, hemoglobin would remain approximately 90 percent saturated.

**If hemoglobin is saturated even at low pressure, why are some patients given 100 percent oxygen?**
This increases the amount of oxygen dissolved into the plasma and thus increases the amount unloaded to tissues.

At internal respiration, the oxygen gradient overcomes this affinity for hemoglobin and is unloaded to the tissues from the blood by diffusion. For normal activity, approximately 20–22 percent of oxygen is unloaded to the tissues. This may at first seem like a waste, but it is in fact a reserve. Under heavy exercise conditions, up to 80 percent of oxygen may be unloaded to the tissues. This is partly driven by what is called the **Bohr effect**. Increases in $CO_2$, which would occur during heavy exercise as well as higher temperatures, result in a lowering of the affinity of hemoglobin for oxygen, and therefore leads to the increase in oxygen unloaded.

## Carbon Dioxide

Although some deoxyhemoglobin can bind to $CO_2$ and become carbaminohemoglobin, this will only account for a small fraction of the transported

$CO_2$ in the blood. The majority (70 percent) of $CO_2$ is transported in the blood stream as bicarbonate ions that are dissolved into the plasma.

Carbon dioxide enters the blood stream via diffusion and enters the red blood cells. Under conditions of high $CO_2$ levels, such as what exists at the tissue level during internal respiration, carbonic anhydrase (an enzyme inside of the red blood cells) facilitates the conversation of $CO_2$ into carbonic acid, which rapidly and spontaneously dissociates into hydrogen and bicarbonate ions. Some of the $H^+$ ions bind to hemoglobin, while others are transported into the plasma where they will cause a decrease in blood pH. The bicarbonate ions will also be transported outside of the cell in a process called the chloride shift. During this process, bicarbonate moves outward while the chloride ion is transported inward to offset the charge difference created by transporting of bicarbonate.

Once the blood returns to the lungs and during external respiration, this entire process is reversed (reverse chloride shift). Chloride is transported out of the red blood cells and into the plasma, which allows for the transportation of bicarbonate into the cell. Bicarbonate is converted back into $CO_2$ (facilitated by carbonic anhydrase), which then diffuses out of the blood and into the alveoli.

## Diseases and Disorders

Any problem that reduces the efficiency or the ease with which gases can be exchanged between the blood and the lungs will wreak havoc on all systems of the body. **Hypoxia**, reduced oxygen levels in the blood, will lead to increased respiratory and cardiac rates in an attempt to compensate for this condition. However, this will require more energy and more oxygen, and cannot be maintained as a normal level of physiological function.

### Respiratory Distress Syndrome

**Respiratory distress syndrome (RDS)** occurs in newborns, especially premature infants, due to an insufficient production of surfactant. Type II pneumocytes will begin producing substantial levels of surfactant in the last few weeks of pregnancy. If babies are born too early, before surfactant levels are optimal, the alveoli will be unable to remain open and much less air will be able to fill this level of the respiratory zone where the bulk of external respiration occurs.

To treat these infants, exogenous (artificial) surfactant is administered into the airway and lungs of the infant in order to open as many alveoli as possible. This continues until the baby is capable of making enough surfactant.

## Asthma

This is a chronic obstructive disorder that differs from COPD (see later section) because asthma is reversible. Triggered by either allergic or inflammatory immune mediators, the smooth muscle of the bronchioles will spasmodically and involuntarily constrict and reduce the flow of air into the lungs. Wheezing and shortness of breath result, but can be reversed by blocking beta-2 adrenergic receptors in the airway, thus mimicking an autonomic nervous system response that dilates the bronchioles.

## Emphysema

Emphysema is a nonreversible destruction of lung tissue, specifically alveoli. Any damage, environmental or otherwise, that causes the death of type I pneumocytes and their replacement by connective tissue cells (scar tissue) will reduce the efficacy of the lung tissue. The symptoms will be similar to those of other respiratory disorders and, unlike asthma, are not reversible.

## Chronic Obstructive Pulmonary Disease (COPD)

COPD has claimed an increasing number of lives in Westernized cultures as more and more people are smoking tobacco (the number one cause of COPD worldwide). This disease is somewhat of a combination of asthma and emphysema. It begins with lung irritation, primarily from smoke, and leads to an immune response and inflammation. Chronic exposure of the lung tissue to the harsh chemicals in the smoke will result in chronic inflammation and a reflex bronchospasm response. However, unlike asthma, these constrictions of the airway are poorly reversible and rather permanent in nature. If the environmental stressors are not eliminated, the reduced airflow will lead also to the destruction of alveolar tissue (emphysema). Together the asthma-like reduced airflow and emphysema compose the major problems involved in COPD. Because this disease does not respond well to treatment, avoiding those environmental agents that can lead to COPD is recommended.

CHAPTER 16

# Endocrine System

In addition to the central nervous system, the endocrine system controls many of the functions of the human body, including but not limited to metabolism, reproduction, growth, and activity level. Unlike other exocrine glands, which transport their products via ducts to a specific and discrete location, the glands of the endocrine system secrete their products directly into a capillary bed. Therefore, they utilize the circulatory system to distribute their chemical signals throughout the body. While this is a rapid systemic way to influence the entire body, only those cells and tissues that express the receptors for the signals will respond. Therefore, this system relies on the signals and receptors to initiate a cellular reaction.

# Hormones

The chemical mediators of the endocrine system are collectively termed hormones and are distributed throughout the body via the circulatory system. The intensity of a physiological response depends on the concentration of hormones produced by glands and the density of receptors expressed on the cell surface of the appropriate tissues. If, for example, no receptors are available to bind to even a high level of hormone, no cellular response is initiated. This means that some tissues will never respond to a particular hormone because they do not have the receptors for that hormone, while other tissues may temporarily decrease the number of receptors, so, for the short term, they will be unable to be stimulated by the hormone. Any substance that is transported in the blood and that elicits a cellular response may be termed a hormone.

## Amino Acid Derivatives

Amino acids are the building blocks (monomers) of proteins. However, individual amino acids may be transformed into other biologically significant molecules such as hormones. This is the case for the amino acid tyrosine. This amino acid is the starting material for a group of hormones called the **catecholamines** (a specific monoamine derivative of tyrosine). The amino acid phenylalanine is converted into tyrosine, which may be further processed into dopamine. This catecholamine functions as either a hormone or a neurotransmitter depending on where it is produced and how it is transported. In the brain, dopamine is essential for motor control, as well as for mood associated with reward and gratification. Outside of the nervous system, dopamine is a potent vasodilator and regulator of another catecholamine (norepinephrine). Interestingly, this is the next phase in the processing of tyrosine. Dopamine is first converted into norepinephrine before it is processed into an essential body hormone, epinephrine (adrenaline). Additionally, tyrosine may be processed by thyroid cells into a differently iodinated form, either **thyroxine** or **triiodothyronine**.

Tryptophan is another amino acid that may be processed into hormones. Serotonin and melatonin are derived from tryptophan in the pineal gland and play a role in diurnal (sleep/wake) cycles.

### Proteins

The largest molecular group of hormones is the protein hormones. Many of these are produced in the pituitary gland (master endocrine gland) and control functions from water retention in the kidneys (e.g., antidiuretic hormone, ADH), thyroid hormone secretion (i.e., thyroid-stimulating hormone, TSH), and reproduction (luteinizing hormone, LH, and follicle-stimulating hormone, FSH).

**Pancreatic islet cells** also produce and secrete protein hormones such as insulin and glucagon, which regulate the carbohydrate level in the blood stream. Other protein hormones (i.e., calcitonin and parathormone) regulate the level of calcium in the blood stream.

### Steroids

While cholesterols are often viewed in a negative light, without them the human body would not have steroid hormones. Cholesterol is the starting material upon which the steroid hormones are based. These include testosterone, estrogen, and progesterone, which function in human reproduction. Cortisol (cortisone) is another steroid hormone produced in the adrenal gland that takes part in many functions of metabolism, especially carbohydrate release into the blood stream.

## Pituitary Gland

The **pituitary gland** (hypophysis) is the master regulatory gland of the human body. This endocrine gland produces and secretes hormones into the blood stream that will lead to the release of other endocrine hormones from distant glands.

### Anatomy

The pituitary gland is found at the base of the brain and suspended by a stalk (**hypophyseal stalk**) just inferior to the hypothalamus. Only about the size of a pea, the pituitary comprises 2 lobes that are derived from distinctly different embryological tissues. These tissues will also be components of the hypophyseal stalk.

During embryonic development, the floor of the future brain will protrude downward as a small diverticulum called the **infundibulum** gives rise to a portion of the stalk as well as the posterior lobe. Similarly, a portion of the roof of the oral cavity will be projected upward as Rathke's pouch to form the glandular anterior lobe as well as the glandular tissue surrounding the stalk (pars tuberalis).

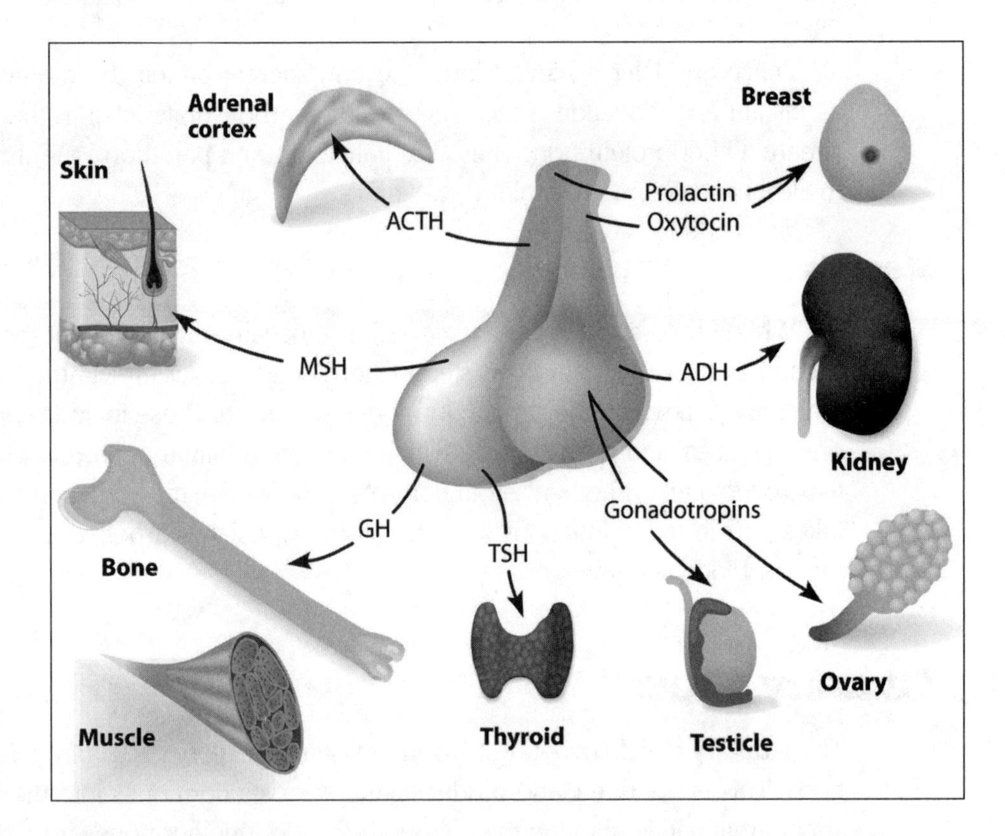


The pituitary gland is the master regulatory gland of the human body. The pituitary produces a number of hormones that have a profound influence on many tissues of the body.

## Anterior Pituitary

The **anterior pituitary** (pars distalis and adenohypophysis) resembles glandular epithelium because of its embryonic source. It is composed of cells that are identified by their staining histologically. Collectively these cells are called

**chromophils** and are subdivided into groups based upon which histological stain colors the cells best.

Acidophils stain a reddish color with eosin and are the most abundant cell type in the anterior pituitary. These reddish cells are further subdivided based upon the hormones they secrete. The **somatotrophs** produce the growth hormone somatotropin, while the **mammotrophs** secrete prolactin. This hormone, stimulated by the pituitary hormone oxytocin, promotes mammary gland development as well as lactation.

**Basophils**, which appear blue in histological preparations, are subdivided into the **corticotrophs** that produce **adrenocorticotropic hormone** (ACTH). This hormone stimulates the cortical cells in the adrenal gland to release other endocrine hormones. **Thyrotrophs**, another basophil cell type, produce and secrete thyroid-stimulating hormone (TSH, thyrotropin). As the name of this hormone implies, it will stimulate the production and release of the thyroid hormones thyroxine and triiodothyronine. The final type of basophil is the **gonadotrophs**, which produce follicle-stimulating hormone (FSH) and luteinizing hormone (LH). In females, these hormones will lead to the maturation of ovarian follicles, ovulation of a mature follicle, and lactation of the breasts at the end of pregnancy. In males, FSH stimulates the development of sperm stem cells (spermatocytes) and LH stimulates testicular cells to produce testosterone.

## Posterior Pituitary

This portion of the pituitary isn't glandular at all. In fact, the **pituicytes** (cells of the posterior pituitary) more closely resemble neuroglia cells than glands. Since this tissue is derived from embryonic forebrain tissue, these cells will support neurons whose axons extend from the hypothalamus to the posterior lobe of the pituitary gland.

## Hypothalamo-Hypophyseal Tract

Hormones produced in the hypothalamus are transported via axons down the hypophyseal stalk and terminate in the posterior pituitary gland. Here they are stored in granules called Herring bodies in the axon terminals. These are the characteristic features of the posterior pituitary gland. The axons make up the hypothalamo-hypophyseal tract, and are responsible for the transporting of hormones and their ultimate and regulated secretion within the posterior pituitary.

The hypothalamic hormones released by the posterior pituitary include antidiuretic hormone (ADH) and oxytocin. ADH (vasopressin) is released when low blood pressure is detected. This hormone will travel to the collecting tubules of the kidneys, where they will increase the expression of **aquaporins** to conserve water and reduce urine production. Greater fluid volume retained in the blood stream will result in increased blood pressure.

Oxytocin is released at the end of pregnancy and functions in a positive feedback mechanism with the smooth muscle of the uterus. Uterine contractions lead to the release of increasing amounts of oxytocin. More oxytocin results in stronger and more frequent contractions of the uterus ending in delivery of the baby and placenta. Additionally, oxytocin has been found to play essential roles in the brains of men and women in the area of intimacy and sexual activity.

**Where does the word "oxytocin" come from?**
In Greek, *oksys* means "swift" and *tokos* means "birth." Thus *oxytocin* means "swift birth."

## Thyroid Gland

The thyroid gland is a bilobed endocrine gland positioned just inferior and ventral to the larynx. Often, the 2 lobes are joined together across the midline of the trachea by a narrow strip (isthmus) of thyroid tissue. Additionally, about half of the population will have a small upward projecting lobe (**pyramidal lobe**) from the midline isthmus upward toward the larynx.

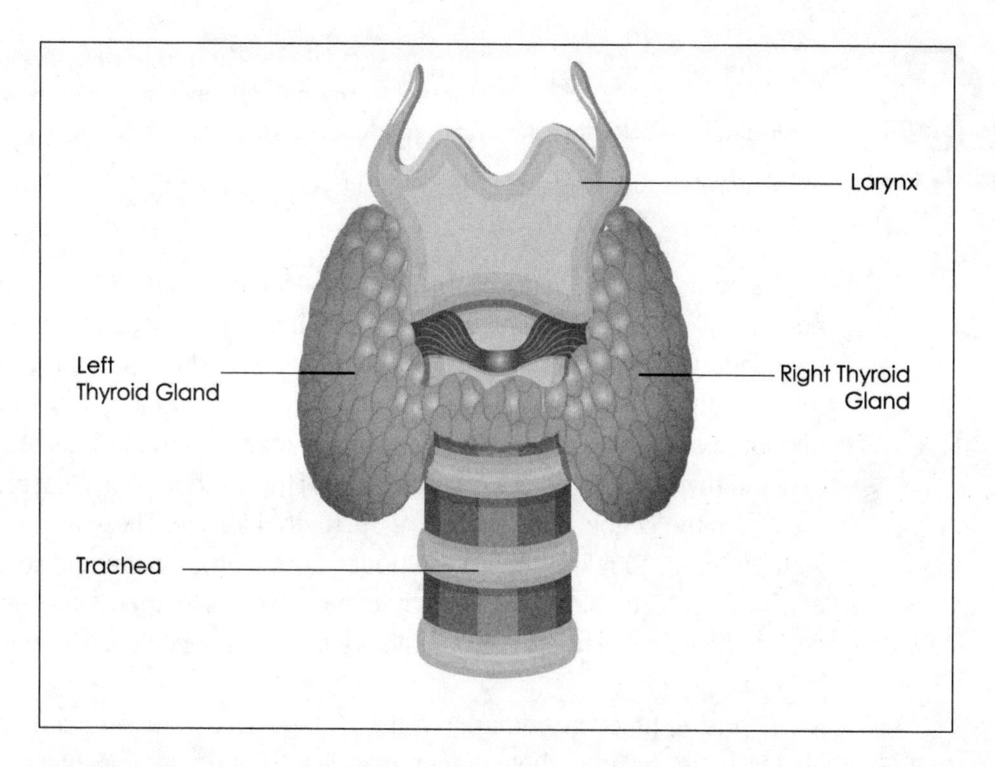


This figure illustrates the shape and location of the thyroid gland on the ventral surface of the neck in the region of the voice box.

## Thyroid Follicles

The thyroid tissue is organized into pools of hormones and stabilizing proteins that are stored outside the cell as colloid bounded by a sphere of thyroid cells (**follicular cells**). They make up the basic unit of the thyroid: the thyroid follicle. The function of the follicular cells is to produce, store, and release thyroid hormones upon the stimulation of the thyroid cells by the pituitary-derived TSH.

## Hormones

Follicular cells begin their production of thyroid hormone with a base glycoprotein called **thyroglobulin**. This protein is rich in the amino acid tyrosine. Cells will add iodine onto tyrosine.

Iodine dietary deficiency will lead to thyroid dysfunction, which is why iodine was added to American-made table salt in the mid-1920s.

Using **thyroperoxidase** enzymes (enzymes that will link iodine to protein), first a single iodine (forming **monoiodotyrosine**) then a second iodine (forming **diiodotyrosine**) will be added to the tyrosine residue. Next, proteases will release either the singly or doubly iodinated tyrosine and add it to an existing diiodotyrosine, creating the final forms of thyroid hormone. **Triiodothyronine** ($T_3$) and **thyroxine** ($T_4$) are the active forms of thyroid hormones that are recovered from the colloid in the center of the thyroid follicle. These are processed with proteases to release these hormones from thyroblobulin and secrete the hormones into the blood stream. Thyroxine amounts to approximately 90 percent of the released hormone, but $T_3$ has a much higher efficacy for physiological stimulation.

$T_3$ and $T_4$ function primarily in the metabolism. Once secreted, they will lead to increases in carbohydrate metabolism, heart rate, appetite, and respiration. At the same time, they will decrease the production of cholesterol and triglycerides and aid in the reduction of body weight.

Hyperthyroidism and hypothyroidism, which may arise due to genetics or from autoimmune disorders that render the thyroid ineffective, lead to the inability to gain weight or lose weight, respectively.

## Parafollicular Cells

In addition to the thyroid follicular cells, other cells reside just on the periphery of the follicles. These cells are seen as aggregates of larger, more rounded cells than the cube-shaped follicular cells. These are the **parafollicular cells**. Parafollicular cells take advantage of the fact that the thyroid is dense with capillaries, since it is an endocrine gland, by secreting their hormones into the same capillaries that the follicular cells are using to transport the thyroid hormones.

Parafollicular cells produce and secrete **calcitonin** (hormone that lowers blood calcium) in response to high blood calcium levels. Calcitonin inhibits bone reabsorption, thereby leaving more calcium stored in the matrix of the bone and reducing the free circulating levels of calcium in the blood stream. Therefore, parafollicular cells are often referred to as C cells.

## Parathyroid Glands

Present on the posterior portion of each lobe of the thyroid gland, the parathyroid glands exists in 4 clusters (superior and inferior portions on both the right and left lobes). The primary cell of this gland is the chief cell. Endocrine in nature, the chief cells produce **parathormone (PTH)**, which functions in an antagonistic fashion with calcitonin. Where calcitonin lowers calcium levels in the blood, PTH increases blood calcium levels. It does so by stimulating bone reabsorption, preventing calcium loss in the urine, and increasing vitamin D production in the kidneys. Vitamin D will function in the alimentary canal to more effectively absorb calcium through the intestinal wall.

Since the names are similar, it is easy to confuse parafollicular cells and parathyroid glands. The former surrounds the follicles (inside the thyroid) and the latter is outside of the thyroid gland.

## Adrenal Glands

Best known for its production of adrenaline under stressful conditions, the adrenal gland is positioned as a cap on the superior portion of each kidney. This gland is often referred to as the **suprarenal gland**. The interior of the gland is much like what you find when you slice open a melon. There is an outer portion, much like the rind of the melon (cortex of the gland) and the sweet part in the interior of the melon (the medulla of the gland). Each of these regions is specialized by cells and the hormones that they produce.

## Adrenal Cortex

Making up approximately 80–90 percent of the mass of the adrenal gland, the outer cortex can be divided into 3 distinct and functionally diverse zones.

### Zona Glomerulosa

The most external of the cortical layers is the **zona glomerulosa**. Its cells are arranged in spherical structures resembling a glomerulus (globular structures of entwined vessels or fibers). The principal function of these cells is the production of mineralocorticoids (hormones that control the retention of water by the kidneys), such as aldosterone. Once secreted, aldosterone will cause the tubules of the kidneys to absorb more sodium and secrete potassium. The result will be the increased conservation of water and a reduction in urine volume.

### Zona Fasciculata

The middle cortical layer, the zona fasciculata, is composed of rows of cells whose cytoplasm is filled with vesicles that give it a rather spongy appearance. These cells are called **spongiocytes**, and they produce glucocorticoids (hormones that control carbohydrate metabolism), such as cortisol. These hormones control general metabolism and have both anabolic (building up) and catabolic (tearing down) effects. For example, cortisol will promote the uptake and storage of fatty acids, amino acids, and glucose. Additionally, it will lead to lipolysis in fat cells (the breakdown of fat) and the breakdown of the protein in muscles that precedes their remodeling.

### Zona Reticularis

The smallest layer and the one adjacent to the adrenal medulla is the **zona reticularis**. These cells produce androgens that will produce a weakly masculinizing effect.

## Adrenal Medulla

The darkly stained interior of the adrenal gland is the **medulla**. The medulla is populated almost exclusively by large, spherical chromaffin cells, and is the site of adrenaline (epinephrine) production. While this is the bulk of what is secreted by the chromaffin cells, about 15 percent of their secretion is also norepinephrine, which is necessary to convert tyrosine into adrenaline.

## Pancreatic Islets

Named because they look like islands of endocrine cells surrounded by the exocrine cells of the pancreas, the cells of the **pancreatic islets** produce hormones that relate to metabolism, digestion, and pancreatic function.

**Alpha cells** in the islets produce glucagon, a hormone secreted when low blood glucose levels are detected. This hormone will result in the glycolysis of glycogen and the addition of glucose to the blood stream, thus increasing blood glucose levels.

**Beta cells** are insulin-producing cells. This pancreatic endocrine hormone is antagonistic to glucagon and will be secreted when high blood glucose levels are detected. In response, cells will remove glucose from the blood, polymerize glucose into glycogen, and store it intracellularly. The liver and muscle cells are well adept at this type of carbohydrate storage.

**Delta cells** produce somatostatin (growth-inhibiting hormone). As the name implies, its job is to regulate the secretion of other endocrine hormones that stimulate growth, and by doing so, slow or halt the growth process in humans.

**Gamma cells** are stimulated after a meal rich in protein, fasting, or exercise. These activities result in a lowering of the blood sugar level and cause the cells to release **pancreatic polypeptide (PP)**. This hormone regulates the pancreas itself, affects glycogen stores in the liver, and stimulates the alimentary canal.

Finally, **epsilon cells** will secrete ghrelin when the stomach is empty so you will feel the sensation of hunger.

## Pineal Gland

This small pinecone-shaped gland (hence the name *pineal*) is located near the center of the hemispheres of the brain in the region of the epithalamus. Often considered the evolutionary derivative of the parietal eye, or "third eye," in lower vertebrates, this gland is involved in the production of melatonin, which is known to control sleep cycles and circadian rhythms.

## Diseases and Disorders

With such diverse chemical regulators controlling almost every function of the human body, any significant glandular problem has severe to dire consequences for the individual.

### Gigantism and Acromegaly

Both of these conditions arise from an overactive pituitary gland, or **hyperpituitarism**. This increase in hormone production, especially those hormones involved in growth, will lead to exaggerated features of the human body that may result in overall larger proportions (gigantism) or isolated enlarged body parts (acromegaly).

If the pituitary gland is active because of a congenital or developmental irregularity causing it to overproduce growth hormone throughout the lifetime of an individual, that person will display the characteristics of gigantism. However, if the pituitary becomes hyperactive after puberty, especially when the growth plates of the long bones have ossified (become bone), only those body parts capable of further growth will show the increases in size. In this stage of life, the disease is termed acromegaly and will display overgrowth of body parts including the hands, feet, lower jaw, and brow ridge of the forehead.

### Graves' Disease

Just as an overactive pituitary can cause the condition of gigantism, an overactive thyroid gland will cause Graves' disease. This autoimmune disease will lead to the doubling in size of the thyroid (goiter), hyperthyroidism, and an increase in all of the physiological activities controlled by thyroid hormones. Individuals with Graves' disease will suffer from hypertension (stemming from increased heart rate), insomnia, weight loss (due to increased metabolism), and extreme fatigue and muscle weakness. However, the most classical sign of Graves' disease is bulging eyes (exophthalmos).

### Diabetes

Diabetes (diabetes mellitus) results from either a lack of insulin production or failure of insulin receptors to detect and respond to secreted insulin. Regardless of the molecular defect, the physiological result is an inability to reduce sugar levels in the blood stream. As glucose levels increase in the blood (hyperglycemia), glucose will begin to be excreted in the urine (glycosuria) and will eventually rise to such levels as to render the individual unconscious. Symptoms of diabetes are not unlike those seen in hyperthyroidism, which include weight loss and increased appetite. However, more frequent urination and increased urine volume will distinguish diabetes mellitus from any thyroid condition. Close monitoring of blood glucose levels, along with manual administration of insulin, are ways an individual can cope with this disease.

## Diabetes Insipidus

Although termed diabetes (which is derived from the Greek word for "siphon"), this condition has nothing to do with glucose or carbohydrate levels. Both conditions do, however, lead to the production of increased amounts of urine—hence the name. Diabetes insipidus is caused by a lack of production of ADH by the pituitary gland or failure of ADH receptors in the kidney tubules to detect and respond to ADH. Without ADH, the aquaporins (protein channels through which water can more easily diffuse) in the collecting ducts of the kidneys will not be expressed, and therefore water reabsorption cannot be regulated. This leads to the production of what many describe as copious amounts of diluted urine. Typically, an individual will produce approximately 1.5 L of urine per day. However, a person suffering from DI will produce in excess of 3 L and possibly up to 15 L of urine per day.

CHAPTER 17

# Urinary System

The principal function of the urinary system is to remove toxins from the plasma of the blood and eliminate them from the body in the form of urine. In doing so, the kidneys will filter a huge volume of plasma from the blood daily (around 180 L of filtrate will be produced per day). However, all but approximately 1.5 L of fluid will be returned to the body; the remainder will become urine. The kidneys play an important role in homeostasis, which is fluid retention (reabsorption). Depending upon the volume of urine produced versus fluid retained, the kidneys will also function to regulate blood pressure by altering the fluid volume of the blood. Additionally, the kidneys will reabsorb all of the essential components from the filtrate and return them to the blood stream. Among those elements that will be completely returned to the blood are proteins and carbohydrates (glucose). All things considered, the kidneys actually are organs of conservation that also play a minor, yet critical, role in toxin elimination.

## Kidneys

The bilateral kidneys are bean-shaped organs located in the lower lumbar region of the abdomen. The kidneys receive a supply of blood from a single renal artery and a single renal vein returns the blood to the inferior vena cava. A continuous supply of blood is essential for kidney function, because urine production begins with filtered blood plasma entering the kidney tubules where the plasma will be modified, concentrated, and excreted as urine.

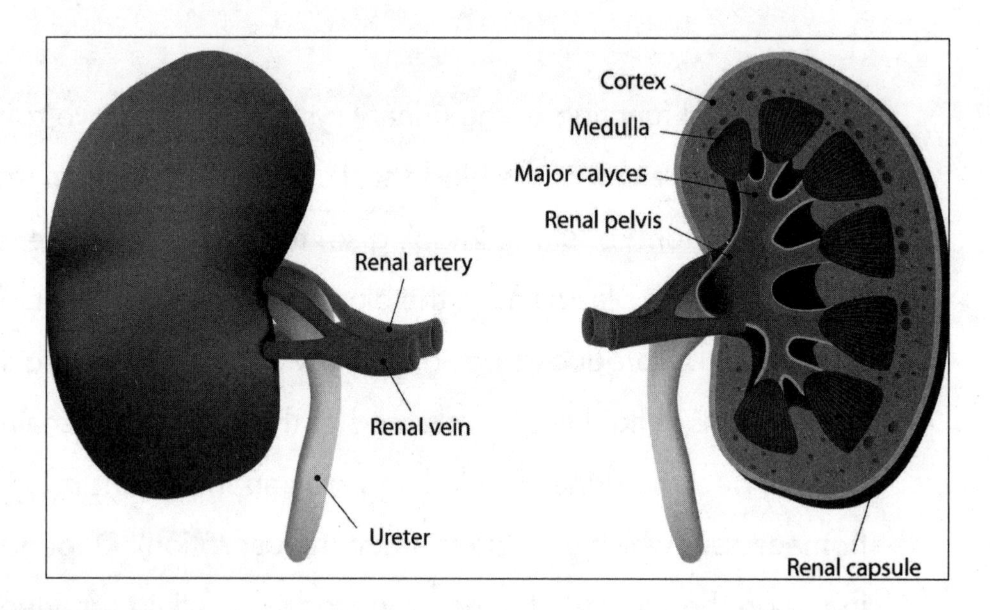


Urine is conveyed from the kidney via the ureter.

## Anatomy

The kidney has an indention (**hilum**) facing the midline of the body and a convex surface facing the lateral part of the abdominal cavity. At the hilum, blood vessels will enter and leave, as does the ureter, which is the tube that transfers urine from the kidney to the bladder. With 2 kidneys servicing the body, an individual may lose 1 kidney and the remaining organ will be sufficient for survival.

## Flow

Upon internal examination of the kidney, one will see the kidney organized into an outer cortex and an inner medulla, much like that of other organs, such as the adrenal gland. Both regions will contain the vascular elements that will supply the organ as well as be used to provide the plasma for urine production. From the renal artery, **interlobar arteries** will branch and extend upward through the medulla until the boundary of the medulla and cortex is reached. At this point (the **juxtamedullary zone**), **arcuate arteries** will branch and follow the curvature of this zone, which will mimic the convex curvature of the kidney itself. Perpendicular branches from the arcuate artery will course upward into the cortex as interlobular arteries. Here, branches called **afferent glomerular arterioles** will arise and provide blood to the capillary bundle called the glomerulus where the plasma will be filtered from the blood and urine production will begin.

While in a typical circulatory system venules follow capillaries, in the kidney the efferent glomerular arteriole follows the capillaries of the glomerulus. Although plasma is lost, no gases have effectively been exchanged sufficiently to reduce the oxygen content to that of the venous system. These "exiting" arterioles will now supply a network of **peritubular capillaries**, which are capillaries that will carry out internal respiration for all the cells of the kidneys. This network will begin in the cortical region of the kidney as long, straight extensions of vessels coursing first downward into the medulla as the arteria recta (straight arterioles) before returning upward into the cortex via the vena recta (straight venules). These 2 straight vessels compose the **vasa recta** component of the kidneys and will extend peritubular capillaries between the 2 straight vessels, hence the conversion from arteriole to venule.

The venous blood is now drained into an interlobular vein with the remaining vessels removing blood from the kidneys mirroring their arterial counterparts (arcuate veins, interlobular veins, and the renal vein).

## Renal Pyramids

The interior of the kidney is also divided into units called **renal pyramids**. Essentially, the tubules and cells of the kidney are arranged as triangles with the apex of each pointing downward toward the hilum. In this way, as urine is produced, it will be moved through a series of funnel-like units to this single point, where it will leave the kidney via the ureter. Each of the 10–12 pyramids will end at its apices (**renal papilla**) and empty the urine into the first of 3 cone-shaped funnels called a **minor calyx**. Several minor calyces then empty into between 3–5 larger major **calyces** that will all be drained by a single large funnel, the **renal pelvis**, which funnels urine to the ureter.

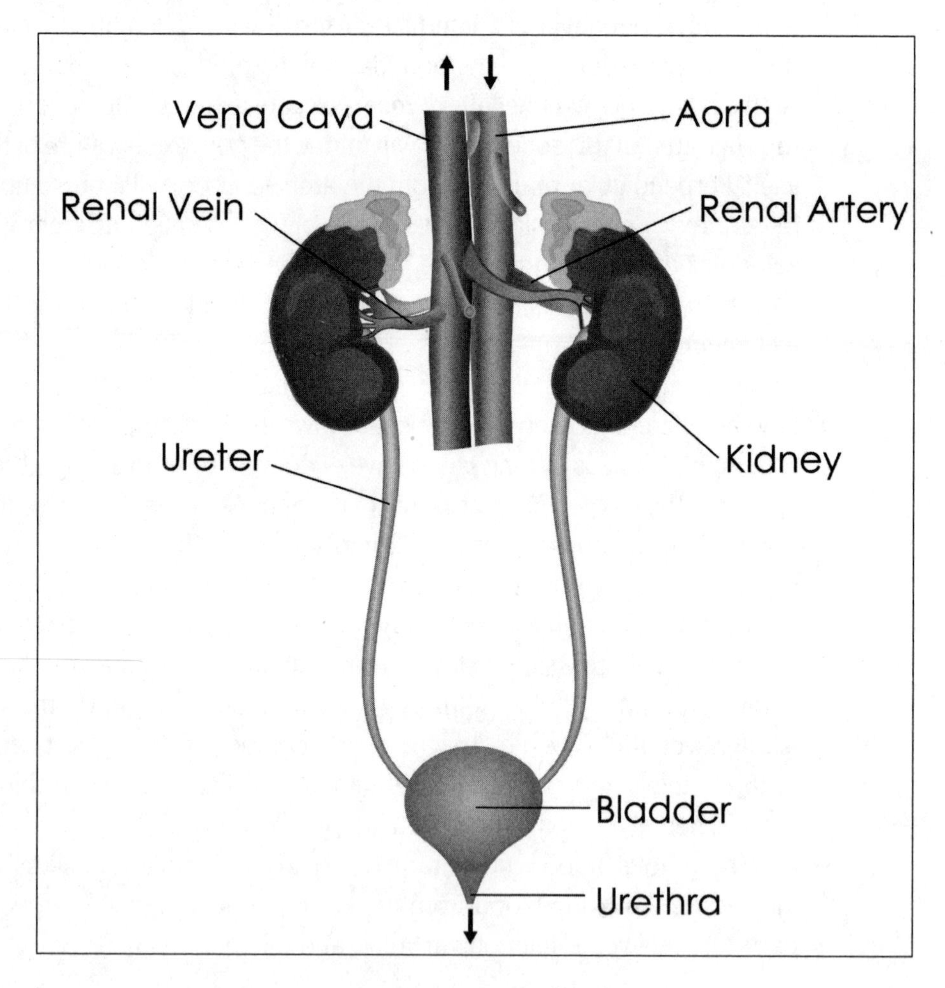


Urine is produced in the kidneys, transferred to the bladder via the ureters, and finally eliminated from the body via the urethra.

# Nephrons

Nephrons are the network of tubes that will transport and modify the plasma-derived filtrate into urine as it courses its way through both the cortex and medulla of the kidneys. At the beginning of the tubules is a capsule that will completely surround the glomerular capillaries and capture the filtered plasma as it leaves the vessels and enters the tubules. The following sections detail the specific regional distinctions that collectively compose the nephron of the kidney.

## Renal Corpuscle

The glomerulus is the cluster of capillaries where plasma from the blood will leave and enter the initial space of the **nephron**. Bowman's capsule (**glomerular capsule**) surrounds the glomerulus to trap the filtered fluid in the space between the inner (visceral) layer and the outer (parietal) layer of the capsule. You can envision this arrangement by taking your fist and placing it in the palm of your other hand with your fingers over the closed fist. The fist is similar to the glomerulus and the covering hand and fingers represent Bowman's capsule. Together, the glomerulus and its associated capsule are termed a **renal corpuscle** and will only be found in the cortex of the kidney.

## Proximal Convoluted Tubule

Glomerular filtrate will flow through Bowman's space and into the beginnings of the **proximal convoluted tubule (PCT)** at the urinary pole of the renal corpuscle. This section of the nephron is also restricted to the cortex of the kidney, much like the renal corpuscle. The cells within the PCT are well suited to start the process of reabsorbing essential materials back into the blood stream from the filtrate. All glucose and all proteins, under normal physiological conditions, will be reabsorbed by the time the filtrate enters the next section of the tubular network.

Additionally, 65 percent of the sodium chloride (NaCl) and water in the filtrate will be actively and consistently reabsorbed in the PCT. This occurs in a constitutive manner and cannot be altered. The PCT cells reabsorb $Na^+$ using the active transport $Na^+/K^+$ pumps. Chloride ions ($Cl^-$) will be reabsorbed by diffusion, and the electrostatic attraction between $Na^+$ and $Cl^-$ will result in NaCl being enriched in the tissue surrounding the tubules. Therefore, the salt creates a hypertonic environment (greater solute concentration) that will generate an osmotic pressure and draw water out of the filtrate through the PCT cells.

Filtration is the process where materials leave the blood stream. Absorption (reabsorption) is the process of returning materials to the blood stream.

## Intermediate Loop

After the PCT, the modified filtrate begins to descend from the cortex into the medulla in the first portion of a **medullary loop**, termed the **intermediate loop** (the loop of Henle). This is called the intermediate loop because it is between the PCT and the next cortical portion of the nephron, that being the **distal convoluted tubule (DCT)**. The first portion of the loop is to the PCT, and is therefore called the **thick descending limb** of the loop. From this region, another 20 percent of the salt and water will be reabsorbed using the same mechanisms as in the PCT. By this point, of the 180 L of filtrate that is produced per day, 27 L remain in the filtrate, which may be reabsorbed under hormonal control down to the approximate 1.5 L of urine that will be eliminated.

Diving deeper into the medulla, you will find a much narrower section of the loop, appropriately termed the **thin descending limb**, and, following a curve, a **thin ascending limb**. These regions will be actively involved in water reabsorption. Following a **thick ascending limb** of the loop, which resembles the next tubular section histologically and partly functionally, the filtrate enters the DCT.

## Distal Convoluted Tubule

Like the renal corpuscle and PCT, the DCT is found in the cortex of the kidney and transports filtrate that has been processed through the earlier regions of the nephron. This region does little for the reabsorption of water. However, it does play an essential role in the acid-base balance of the urine, and in doing so, affects the pH of the blood.

### Collecting Duct

Having left the DCT, the last segment of the nephron, the filtrate now enters a collecting duct with a larger diameter. Several nephrons will empty their filtrate into a collecting duct and will begin the process of funneling the filtrate (and urine) toward the renal pelvis and ureter. As the filtrate moves down the collecting duct, the diameter will slowly increase as the duct approaches a renal papilla of one of the pyramids. The ducts are not termed papillary ducts (a.k.a. ducts of Bellini) and will leave the papilla and enter the space of a minor calyx via the area cribrosa of the papilla.

Although not a portion of a nephron, the collecting duct along with a nephron is considered the functional unit of the kidney and is called a **uriniferous tubule.**

## Ureter, Bladder, and Urethra

From the kidneys, which produce urine, through the ureter to the bladder, where urine is stored, then through the urethra to the exterior of the body, this modified plasma has made its way through the entirety of the urinary tract.

### Ureter

This muscular tube interconnects the kidney with the bladder. Much like in the alimentary canal, the smooth muscle of the ureter will spasmodically contract in peristaltic movements to propel urine toward and into the bladder. Only 3–4 mm in diameter, this is typically the location where one becomes aware of any kidney stones that are being passed through the urinary tract. Although the lining of the ureter is transitional epithelium, which can stretch as the tract fills with urine, it will not expand to the extent that most stones need to pass freely.

### Bladder

The urinary bladder receives urine from each kidney via the ureters and can store approximately 500 ml–1 L of urine. Lining the bladder is the same compliant transitional epithelium found in the ureter, as well as underlying layers of smooth muscle that collectively are referred to as the **detrusor muscle**. These layers are under autonomic nervous system control and when stretched reflexively contract in order to urinate (**micturition**).

### Urethra

At the inferior apex of the bladder is the opening of the **urethra**, the tube that will transfer urine from the bladder to the outside of the body. It is in this region that you would find the internal urethral sphincter, which prevents urine from leaking into the urethra. As the urethra passes through the perineal muscle of the lower pelvis, the skeletal muscle creates the external urethral sphincter. Unlike the internal, the external sphincter is under voluntary control to prevent early voiding of the bladder. In newborns, this is a learned process; it takes time to overcome the involuntary opening of the internal sphincter upon stretching and the sensation of "urgency."

Much shorter in females than in males, this conduit for urine has much less smooth muscle than the ureter, and will shift from transitional epithelium near the bladder to stratified squamous epithelium (similar to the skin) near the external urethral orifice (opening at the end of the urethra).

## Urine Formation

Urine production begins when plasma leaves the glomerulus and enters Bowman's space. Throughout the nephron and the collecting duct, materials will be removed from the filtrate (reabsorption) or may be added to the filtrate via the tubule cells themselves (secretion) in order to conserve essential materials and eliminate those in excess, and to balance blood levels and remove toxins.

A diuretic is a substance that, when consumed, increases the production of urine. Water is a diuretic, because the more water you drink, the more urine you produce.

## Glomerular Filtrate

**Glomerular filtrate** is often termed ultrafiltrate since it is not exactly the same as the plasma found in the blood. In fact, plasma will pass through 3 filtration mechanisms before it is able to enter into Bowman's space.

## Glomerular Capillaries

Glomerular capillaries are a fenestrated type of capillary. This means that rather than materials diffusing first into the cell then diffusing out the other side, materials may pass through the pore in the endothelial cell for easier exiting of the blood stream. These pores are called **fenestrations** and are approximately 70–90 nm (nanometers) in diameter. You can think of these as similar to a slice of Swiss cheese. Therefore, plasma will be able to freely flow through these fenestrations with only the formed elements such as RBCs, WBCs, and platelets being retained in the blood vessels.

## Basal Lamina

Underlying every epithelial cell is a molecule-rich layer called the **basal lamina**, and in this case the molecules beneath the fenestrated glomerular capillaries will create a molecular filter for the forming filtrate. The basal lamina is divided into 2 regions. Adjacent to the basal membrane of the cell is a less dense portion termed the **lamina rara**. Here, you will find an abundance of adhesive glycoproteins such as **laminin** and **fibronectin**, which provide anchorage points for the cells to attach to this extracellular foundation. Also found in the lamina rara are **heparan sulfate proteoglycans**. These consist of long linear protein "cores." Polymers of disaccharides called glycosaminoglycans will attach to this core. In this case, the disaccharide is a heparan sulfate. This complexity is essentially geared to localize a high concentration of sulfate groups (negatively charged) in this layer to trap positively charged material from the plasma as it is becoming filtrate. Therefore, the lamina rara presents an ionic filtration step in filtrate formation. Also, since this layer is adjacent to the glomerulus, it is more specifically termed the **lamina rara interna**.

Below this layer is the much thicker and darker **lamina densa**. The greater density of this layer compared to the rara is due to the presence of type IV collagen, which is arranged into a meshwork much like a fishing net. Woven together, the spaces between opposing collagen molecules restrict passage of materials to anything smaller than 69 kD (kiloDaltons) and form a size filtration step. This layer is thicker than the lamina rara interna because it is actually produced by 2 opposing epithelial layers. The capillary endothelial cells are continually producing new basal lamina components and the third filtration step, which is another epithelial layer that is also adding basal lamina components to this layer. This produces another lamina rara, which is below the lamina densa but is adjacent to the outer epithelial layer. This is called the **lamina rara externa**. It is composed of the same components in the lamina rara interna.

Imagine an Oreo cookie. Each cookie is the basement membrane for each epithelium (capillaries and visceral layer). The cookie is the lamina rara, and the cream filling is the lamina densa. If you take two cookies and remove one of the chocolate wafers from both, and then press the cream filling together into a cookie with a lot of filling, this illustrates exactly what happens between these two cell layers. With both of their basal layers facing each other, their basement membranes blend together like the cream filling of the Oreo cookie.

## Visceral Layer of Bowman's Capsule

The cells of this layer are tightly adhered to every loop of the glomerular capillaries and create a functional barrier (filter) for the filtrate to pass through. The modified epithelial cells that compose this layer are called **podocytes** (*pod* meaning "foot") because they have extensive and intricate cellular processes that will completely cover the capillary loops. First, the podocytes extend primary (larger) processes, which extend along the length of the capillary. From these smaller, fingerlike processes called **pedicels** (secondary processes) extend perpendicularly from the primary process and wrap around the circumference of the capillary to interdigitate with the pedicels from adjacent podocytes. This is much like putting your hands together and interlocking your fingers to cup your hands to drink water. In doing so, you realize that spaces still exist between the fingers from which water can leak. Much is the same for the pedicels of the podocytes.

These filtration slits will create a portion of the third filtration step in the renal corpuscle. Additionally, between these slits, a membrane will extend to further regulate the passage of materials into Bowman's space.

## Countercurrent Multiplier

In the intermediate loop, a continuous cycle of salt reabsorption followed by the osmosis of water out of the filtrate is created in the medulla of the kidney. Water is reabsorbed at the level of the descending limb, primarily due to the active pumping of Na (and the passive diffusion of $Cl^-$) that happens in the ascending limb.

Sodium/potassium pumps actively transport $Na^+$ into the **interstitium** (tissue surrounding the loop) and $K^+$ is pumped (secreted) into the filtrate to offset the loss of $Na^+$. Chloride follows the electrostatic attraction of $Na^+$ and diffuses into the medullary interstitium to combine with $Na^+$. This forms NaCl in a mechanism similar to that used in the PCT. It is important to note that the ascending limb is *not* permeable to water, so the filtrate becomes less concentrated with salt, which makes it **hypotonic** (approximately 100 mOsm).

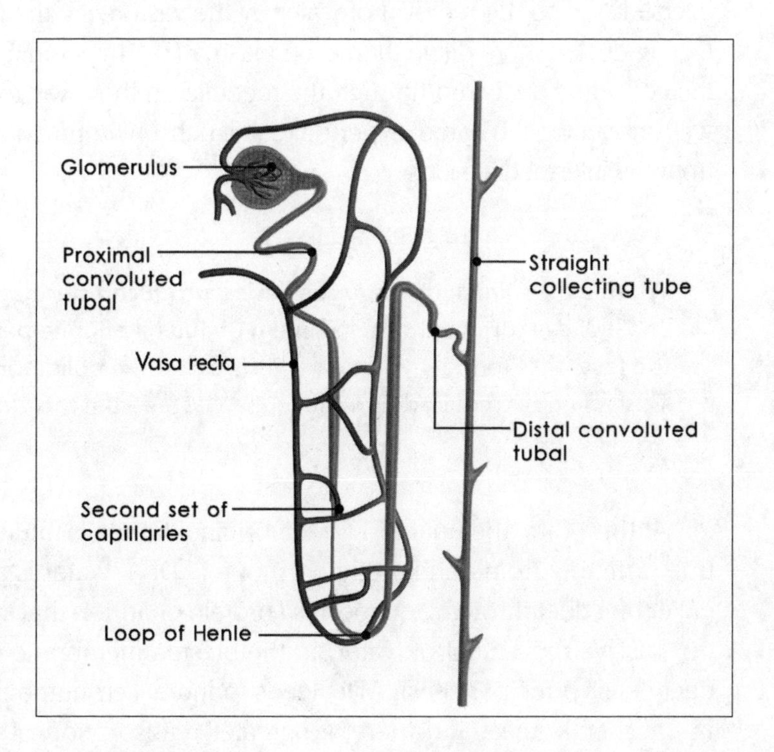


This figure shows the tubular nephron and the associated peritubular capillary network.

Because the NaCl from the ascending loop is accumulating in the interstitium, the deeper levels of the medulla, as well as the filtrate in the descending limb, becomes more hypertonic (1200–1400 mOsm). There are no additional $Na^+/K^+$ pumps in this region of the loop; however, the descending limb is permeable to water, allowing water to be reabsorbed into the interstitium. This causes the filtrate to become more and more hypertonic (1400 mOsm at the bottom of the loop).

As salt and water are reabsorbed into the medulla of the kidney, the vasa recta and peritubular capillaries are responsible for removing the excess salt and water and returning it to the blood stream. If this material were not removed, the kidneys would reach too high a level of salt concentration and be unable to increase further, which would shut down the kidneys. Therefore it is as essential to remove the salt and water from the kidney as it is to remove these materials from the filtrate so that the cycle can continue.

## Collecting Ducts

Due to the active (and passive) movements of solutes into the medullary interstitium and the removal of water by the vasa recta, the filtrate in the collecting duct is more dilute than even plasma (i.e., hypotonic). Remember that the collecting ducts run through the medulla on their way to the calyces and will encounter the same hypertonic, salt-rich environment that drew water from the intermediate loop.

**What is the minimum amount of urine produced per day?**
The kidneys will produce a minimum of 400 ml of urine per day even in the face of severe dehydration. This is called the obligatory water loss. This volume is required to remove the wastes from the blood.

At this point, the final 27 L of the original filtrate is under hormonal control, primarily from antidiuretic hormone (ADH). Water is able to leave the collecting duct through aquaporins (protein channels that specifically allow the passive movement of water) in the plasma membrane of the collecting duct cells (principal cells). ADH leads to increased numbers of aquaporins in the membranes, and therefore greater reabsorption of water. However, even during dehydration, ADH cannot lead to more highly concentrated urine (above that of the interstitial fluid).

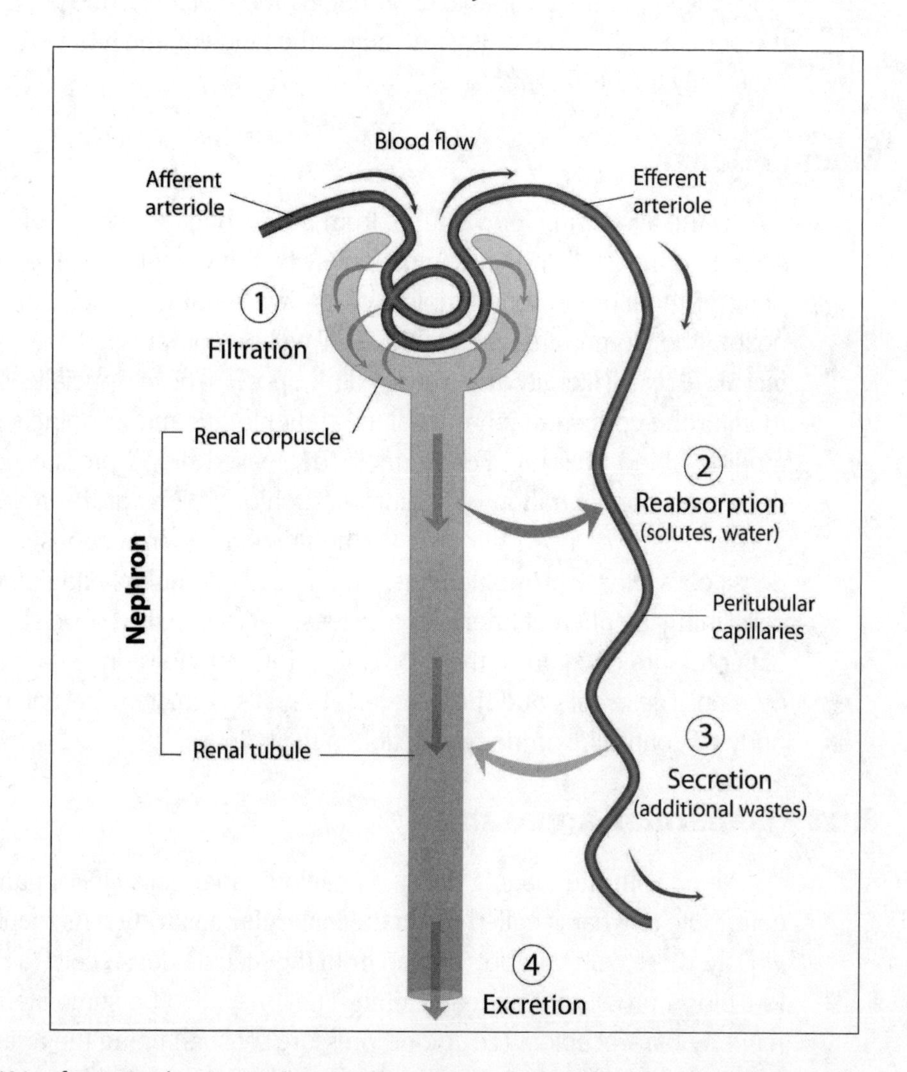


Urine formation begins as plasma is filtered from the blood and moved into the first portion of the nephron. Along the pathway through the remainder of the nephron, the filtrate may have materials removed and returned to the blood stream (a process called reabsorption) or have material added to the filtrate (a process called secretion). Using these three mechanisms, plasma filtrate is converted into urine.

# Body Fluid Balance

Since the kidneys are capable of producing more or less urine per day, they are the perfect organs to monitor and adjust the fluid volume retained in the body, especially the blood stream.

## Macula Densa

As the ascending loop returns from the medulla to the cortex, it will transition into the DCT and pass immediately by the vascular pole (entry and exit point of the glomerular arterioles) of its own renal corpuscle. In this region, next to the glomerulus, cells of the DCT will be compressed together into the **macula densa** (literally translated "dark spot"). These specialized cells will monitor the concentration of sodium and chloride and in doing so indirectly monitor blood pressure. For example, decreased blood pressure results in a decreased concentration of sodium and chloride ions at the macula densa. This is due to reduced filtration by the glomerulus. In response, the macula densa cells release prostaglandins, which trigger granular juxtaglomerular (JG) cells lining the afferent arterioles to release renin into the blood stream. Under high-pressure conditions, these cells signal JG cells to reduce their production of renin. These cells and their chemical signals regulate water retention in the body by controlling urine production in the kidneys.

## Juxtaglomerular Apparatus

Along with the macula densa, juxtaglomerular cells (JG or granular cells) contribute to what is called the **juxtaglomerular apparatus**. As mentioned previously, these cells can be signaled from the macula densa cells (a response to low blood pressure) to release renin. They may also be stimulated to secrete renin by baroreceptors (i.e., blood pressure sensors) lining the arterioles. This is the next step in what is termed the **renin-angiotensin-aldosterone system** (RAAS).

## RAAS System

The renin-angiotensin-aldosterone system (RAAS) is a hormone system that regulates blood pressure and water balance. When blood pressure is low, JG cells secrete renin, which is an enzyme that converts the inactive plasma protein angiotensinogen into angiotensin I. This inactive intermediate will be further processed in the lungs by angiotensin-converting enzyme (ACE) into angiotensin II (active form). Angiotensin II causes blood vessels to constrict, resulting in increased blood pressure. It also stimulates the secretion of aldosterone from the adrenal cortex.

Aldosterone increases thirst and water reabsorption by promoting $Na^+$ reabsorption and $K^+$ secretion in the nephron. Additionally, high $K^+$ (hyperkalemia) directly stimulates aldosterone secretion, resulting in the lowering of blood potassium levels.

# Electrolyte Balance

As with the homeostatic balance of any system in the human body, changes in one area or the concentration of one element may have drastic effects on another. Therefore, as sodium and chloride are reabsorbed (to conserve water) and potassium is secreted to offset the conservation of sodium, an imbalance in electrolytes can easily and quickly occur if not regulated closely. In summary of the previous sections, recall that $Na^+$ regulates blood volume and pressure. However, in addition to being used to balance $Na^+$, potassium is also essential for maintaining proper function of cardiac and skeletal muscle.

## $Na^+$ Reabsorption

Most of the filtered sodium will be reabsorbed in the PCT and intermediate loop (being used to reabsorb water). In the absence of aldosterone, 80 percent of the remaining sodium is reabsorbed by the collecting duct cells. This will result in a daily loss of approximately 30 grams of sodium in the urine. However, in the presence of maximal amounts of aldosterone, no $Na^+$ is lost.

## Potassium Secretion

Like sodium, 90 percent of $K^+$ is reabsorbed in the PCT and intermediate loop. However, secretion of $K^+$ occurs in the late distal tubule and collecting duct. This may actually be in concert with $H^+$ ion secretion, because either $K^+$ or $H^+$ can be used to offset $Na^+$ reabsorption. If blood pH is too low, then more $K^+$ will be secreted. However, in cases of hyper- or hypokalemia, which could lead to cardiac arrhythmias, more $H^+$ will be secreted.

Normally, $K^+$ in the urine matches the $K^+$ ingested because meals rich in $K^+$ lead to aldosterone production and $K^+$ secretion into the filtrate. Diuretics (drugs used to cause an increase in urine production to lower blood volume and reduce blood pressure) that cause $Na^+$ to be reabsorbed at higher levels than normal in the collecting ducts lead to higher $K^+$ secretion and low blood $K^+$ (hypokalemia). Both hyperkalemia or hypokalemia can lead to cardiac arrhythmias.

## Acid-Base Balance

Since $H^+$ can be secreted into the urine, the kidneys play an important role in the maintenance of proper blood, and therefore, body pH. $H^+$ ions are filtered through the glomeruli and may be secreted into the filtrate mechanism with $Na^+$ in the DCT. This could explain the slightly acidic nature of urine. However, the blood contains bicarbonate, which is a pH buffer (prevents drastic shifts in pH).

Recall that pH is a measure of the $H^+$ concentration in a particular solution and is represented on a scale of 1–14, with 7 being neutral. Acids are solutions with high $H^+$ concentration (and lower pH numbers), while bases have fewer $H^+$ (and higher pH numbers).

Unfortunately, much like protein and glucose that finds it way into the filtrate and is completely reabsorbed, so is bicarbonate. The PCT is impermeable to bicarbonate; however, since $H^+$ is secreted, bicarbonate quickly forms carbonic acid. Recall that carbonic acid is formed from $CO_2$ by the enzyme in RBCs, called carbonic anhydrase. This enzyme is also present on the surface of the tubule cells and facilitates the conversion of this spontaneously generated carbonic acid into $CO_2$, which diffuses into the tubule cells. Once inside, an intracellular carbonic anhydrase converts $CO_2$ back into carbonic acid, which dissociates into bicarbonate and diffuses out of the cell and into the vas recta. In this way, 80–90 percent of bicarbonate is reabsorbed. However, in conditions of lower blood pH (higher $H^+$ concentration in the filtrate), PCT cells can produce extra bicarbonate from glutamine, which can be reabsorbed. This is offset with the secretion of ammonia ($NH_3$) and phosphates ($HPO_4^{2-}$), which function as urinary buffers to become $H_2PO_4$ and $NH_4^+$ (ammonium ion). These biochemical mechanisms are attempts to maintain body pH and bicarbonate concentrations in the blood while still being able to remove the nitrogenous waste products from the body in the urine.

## Diseases and Disorders

Any condition that reduces the efficacy of the urinary system will have drastic effects on blood pressure (due to unregulated blood volume) and body toxicity (due to retention of wastes), and may eventually lead to kidney failure and death.

### Kidney Stones

If the kidneys concentrate the urine excessively, then some minerals will become supersaturated and yield a greater chance that through a nucleation event they will begin to form a crystal. As a seed crystal forms, the minerals will accumulate on the crystal and increase in diameter. If the crystal is less than 3 mm in size (the average diameter of the ureter), the stones can be passed in the urine and the individual may not even notice. However, larger stones, which have jagged edges, can become lodged in the ureter until the building pressure of the urine forces the stone down the ureter. This will cause considerable pain as well as damage. Often bleeding from the ureter will appear in the urine (**hematuria**). If the stones become too large to pass safely, sound waves can be used to bombard the stone and shatter it into small enough pieces to pass through the urinary tract. This technique is termed **lithotripsy** (*litho* is derived from the Greek word for "stone"). Other techniques to fragment the stone into smaller pieces include laser catheterization to focus the laser beam directly on the stone.

### Nephritis

Nephritis is inflammation of the kidneys. This is often caused by an infection. Other causes may include increased exposure of the kidneys to toxic agents or an autoimmune reaction that reduces kidney function. Typically, the patient will exhibit reduced urine production, possibly blood in the urine (due to damaged renal corpuscles), and increasing levels of nitrogenous wastes accumulating in the blood (uremia). While infections of the body are not uncommon and most are not thought to be life-threatening, nephritis is a serious disease and is one of the 8 leading causes of human death worldwide.

## Urinary Tract Infection

Any infection of the urinary tract may be termed a urinary tract infection (UTI). Infections in the upper urinary tract are much more serious than those of the lower tract. Most commonly caused by the bacteria of the alimentary canal (E. coli), the infection begins in the urethra and, as the bacteria proliferates, extends upward in the urinary tract. Symptoms of a UTI include, but are not limited to, painful urination (dysuria), cloudy urine due to the excessive number of bacteria populating the urine, an increase in the sensation of urgency, and an increase in frequency of urination. Antibiotics and increased fluid intake (to cause an increase in urine production to flush the bacteria from the urinary tract) are the most common modes of treatment for a lower UTI.

CHAPTER 18

# Male Reproductive System

The true "goal" for any organism is to survive long enough to procreate and ensure the continuation of the species. In human beings, in order to reproduce, a spermatozoon is introduced into the female reproductive tract and is capable of fertilizing an egg. This creates new life.

## Testes

Similar in function as the female ovaries, the testes are where sperm are formed. Here is also where you will find the beginnings of the genital ducts that will eventually transfer sperm from the testes and then to the female reproductive tract.

### Anatomy

Suspended from the **perineum** of the male pelvis, the bilateral testes are covered in skin called the **scrotum**. The scrotum function as more than just a case for the testes; the scrotum and its thin underlying muscles regulate the temperature of the testes by either retracting and pulling the testes closer to the pelvis (to warm the testes) or relaxing and allowing the testes to descend farther away from the body (to lower the temperature of the testes).

Beneath the skin of the scrotum is a dense connective tissue capsule that surrounds each testis called the **tunica albuginea**. This connective tissue will also accumulate at the posterior portion of the testes to form the **mediastinum testis**. From this point, connective tissue septa will emanate throughout the testis to divide the tissue into lobes. Each divided lobe will contain between 1 and 4 seminiferous tubules, which are the sites of sperm germ cells as well as where the sperm will develop.

### Cells

In addition to the germ cells, 2 other cell types are present in the testes that play essential roles in spermatozoa generation as well as the maintenance of secondary male sexual characteristics, which include greater body hair, heavier bone structure and muscle mass, and lower body fat percentage when compared to females, and the development of male genitalia.

**Sertoli cells** (nurse cells) are located throughout the seminiferous tubules and function to sustain and protect the developing sperm. As sperm germ cells undergo meiotic cell divisions and genetic recombination, the resulting sperm will be genetically and immunologically different from the male in which they are produced. If the Sertoli cells did not create a "blood-testis barrier," the newly formed spermatozoa would elicit an immune response and be destroyed. Additionally, these cells produce a fructose-rich secretion that will nourish the sperm in their protected environment within the lumen of the seminiferous tubule.

Outside of the seminiferous tubules, in the interstitium of the testis, are the **interstitial cells of Leydig**. These cells are endocrine in nature and produce the androgen testosterone for the development and maintenance of male secondary sexual characteristics.

## Intratesticular Tract

The pathway sperm use to make their way from the testes to the female reproductive tract is collectively called the male genital ducts or reproductive tract. After the sperm are formed in the seminiferous tubules, they will move toward the mediastinum testis and straight, narrow terminal portions of the seminiferous tubules, which are called the **tubuli recti** (literally translated "straight tubes").

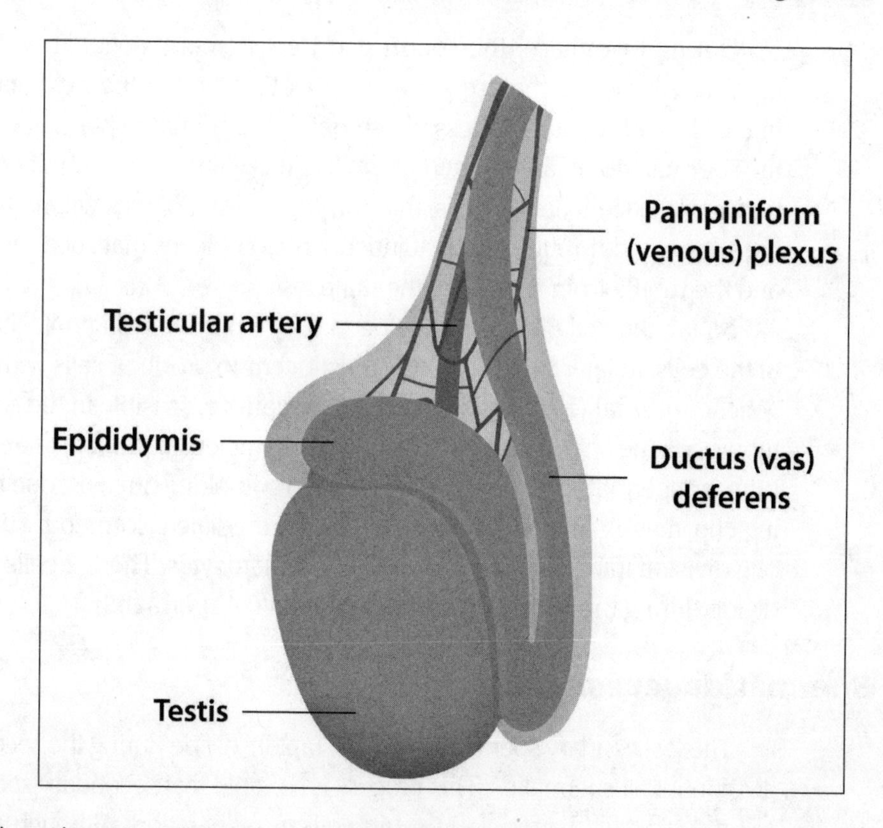


The male testis, including the epididymis, ductus deferens, and the blood vessels that supply these tissues.

These short straight passages allow the spermatozoa to move into an anastomotic (interconnecting) maze of passages in the mediastinum testis, the rete testis. These passages allow the sperm to move into the next portion of the genital duct system, the efferent ductules. There are 10–20 of these ductules that transfer spermatozoa from the testes into the first portion of the extratesticular tract, the **epididymis**.

## Spermatogenesis

Spermatogenesis is the umbrella term that encompasses all of the cellular and molecular processes that change male germ cells (spermatogonia) into a free, mobile spermatozoa.

### Spermatocytogenesis

During this phase, the sperm germ cells (spermatogonia) will divide in order to reproduce themselves as well as to produce the next stage of cells in the developmental process, those being the primary spermatocytes. While the spermatogonia are situated at the **basal region** of the seminiferous tubules between adjacent Sertoli cells, the primary spermatocytes will migrate toward the lumen and through the tight junctional complexes that separate the lumen and the tubule from the rest of the male testes.

So far, the cell divisions have been accomplished using mitosis or cloning of the cells. In later stages, meiosis must occur to produce cells with half of the genetic material (i.e., haploid or 1N). The main cells visible in the seminiferous tubule are the spermatogonia (with its dark nucleus in the basal area) and the primary spermatocyte (with its larger, spongy looking nucleus, due to condensing chromatin). The next stage cellular intermediate is formed by the first meiotic division into 2 haploid secondary spermatocytes. These 2 cells will rapidly divide during the second meiotic division into 4 spermatids.

### Spermatidogenesis

The 2 secondary spermatocytes will rapidly divide during the second meiotic division into 4 spermatids. This process is so rapid that secondary spermatocytes will not typically be visible in a histological preparation. Although meiosis has occurred to transform a single primary spermatocyte into 4 early spermatids, only the nuclei have divided. The cytoplasm of these 4 daughter cells does not separate; therefore, the cells remain interconnected to each other via cytoplasmic bridges.

Early spermatids are rounded cells with no flagellum. These will develop during the later stages of **spermiogenesis**.

## Spermiogenesis

This final stage in sperm formation converts the rounded early spermatid into a late spermatid, which will then be separated from the other daughter cells and released into the lumen of the seminiferous tubule as spermatozoa. Four distinct stages occur to shape and form all of the cellular and molecular components of the spermatozoa prior to their undocking from the Sertoli cell.

The first phase of spermiogenesis is called the Golgi phase. The abundant Golgi apparatuses will produce many hydrolytic enzymes and store them as granules, which will eventually fuse together into one large package called the **acrosome**. Upon contact with the **oocyte** (egg), these enzymes will be released from the sperm and provide a means by which the head of the sperm may fuse with and be inserted into the egg. While this is happening at the head of the sperm, at the opposite side, **centrioles** are forming into a microtubule-organizing unit to produce the base of the flagellum (the **axoneme**), which will also course through the core of the flagellum itself.

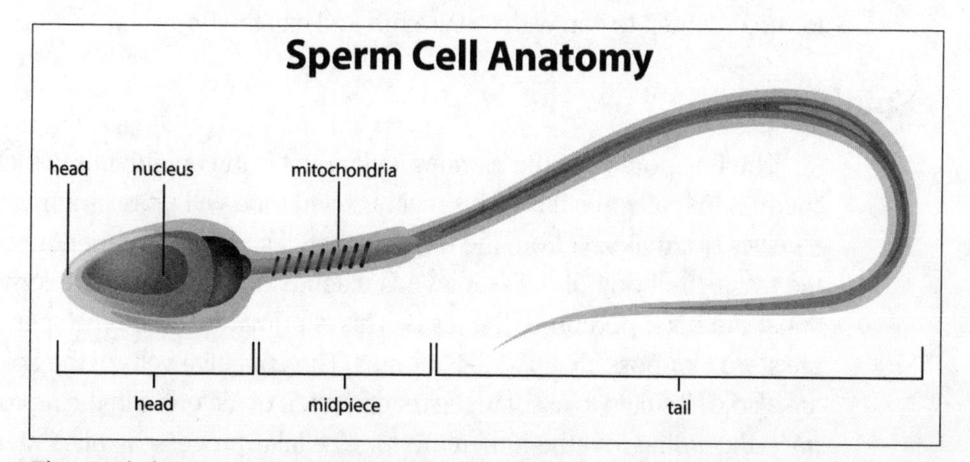


The motile human sperm consists of a head, midpiece, and tail. The head is filled with enzymes used to penetrate the egg, as well a single copy of each human chromosome. The midpiece houses mitochondria that produce energy needed for the movement of the tail. The tail is a flagellum that contains microtubules and associated proteins used to move and propel the sperm through the female reproductive tract.

The cap phase is next. This stage is marked by the movement of all the granules to just above the nucleus of the spermatid, forming the bacrosome (acrosomal cap). Mitochondria are also moved into the area of the flagellum base to provide energy that will drive the movement of the flagellum and propel the spermatozoa.

In the third stage, tail formation, the microtubules extend to push the plasma membrane outward and form the elongated structure that will provide mobility to the spermatozoa. The spermatids at this point are oriented so that their now tapered heads are directed to the lumen of the seminiferous tubule. During the final "maturation" phase, the excess cytoplasm is shed and the spermatids are freed from their sibling cells into independent cells. Any other remaining cytoplasm (or extracellular material) is removed while the spermatozoa are in the lumen of the seminiferous tubule by a process called spermiation (sperm release). However, at this point, spermatozoa are immobile and remain incapable of fertilizing an egg.

## Extratesticular Reproductive Tract

Once the spermatozoa have left the efferent ductules, they enter the **extratesticular ducts**. The first portion of these ducts is found within the scrotum. The later portion is part of the spermatic cord that rises to the pelvis and enters the lower abdomen to join with the urethra and exit the body.

### Epididymis

The first portion of the extratesticular duct is the epididymis, which stores sperm and reabsorbs fluid. This highly coiled tube will possess a head, which receives spermatozoa from the efferent ductules. From the superior portion of the testis, the body of the epididymis extends downward before forming the tail at the back portion of the testis. This is primarily where the spermatozoa are stored for possibly up to 2–3 months. The epithelial cells of the epididymis are also well suited to assist in the reabsorption of materials in the reproductive tract. Extending into the lumen are long cellular processes called **stereocilia**. Stereocilia do not assist in movement of materials; in fact, these processes are actually extremely long microvilli that function by increasing the cellular surface area for maximal reabsorption of material.

## Vas Deferens

Beginning at the tail of the epididymis, the **vas deferens** extends upward from the scrotum in the spermatic cord and enters the body from an opening in the inguinal area (i.e., where the legs join the hips) of the lower pelvis. This is the thickest of the ducts in the male reproductive tract because of its extremely thick layers of smooth muscle. These become active during ejaculation and generate peristaltic contractions that propel spermatozoa along and out of the male reproductive tract.

## Ejaculatory Duct

As the vas deferens approaches the urethra (the common duct for urine and sperm), it is joined by a duct from the accessory sex gland known as the **seminal vesicle**. When this union occurs, this terminal portion of the vas deferens is renamed the **ejaculatory duct**.

Students often incorrectly identify the ejaculatory duct as separate from the vas deferens. It is simply the terminal portion of the vas deferens between the intersection of the seminal vesicle and the urethra.

## Urethra

The final duct of the male reproductive tract is the urethra, which expels urine. Urine is harmful to sperm; therefore, several accessory sex glands are located throughout this portion of the male reproductive tract to not only nourish but also protect the sperm.

### Prostatic Urethra

As soon as the urethra leaves the bladder, it becomes invested by the tissue of the prostate gland (which completely surrounds the urethra). Within this tissue is the ejaculatory duct (and thus the duct of the seminal vesicle). Additionally, **prostatic ducts** will empty materials from the prostate gland (covered in a later section) into the urethra and contribute to the final material that will become the semen.

### Membranous Urethra

This is a short 1–2 cm portion of the urethra that passes through the muscle of the lower pelvic region (the external urethra sphincter). This will also transmit the semen from the abdominal cavity into the last portion of the urethra within the penis.

### Penile Urethra

The longest portion of the urethra is the penile (spongy) urethra. It courses through the ventral portion of the penis, residing within one of the 3 columns of erectile tissue, the **corpus spongiosum** (hence the term spongy urethra). Materials (either urine or semen) will be passed through this duct until the urethral opening (external urethral meatus) is reached, and the materials are expelled outside of the body.

# Glands of the Male Reproductive Tract

Since sperm are essentially a nucleus with half a complement of chromosomes, mitochondria (to provide energy for flagellar movement), and a bag (a.k.a. acrosome) of enzymes to be used to penetrate the oocyte, they have no means of generating or processing materials to use as fuel for the mitochondria. They instead must obtain these materials from the secretions of the accessory sex glands. Additionally, for protection against the acidic urine that may remain in the urethra, other secretions will neutralize and buffer the urethral environment and protect and feed the spermatozoa that are nearing the end of their journey through the male reproductive tract.

## Seminal Vesicles

These accessory glands provide the majority of fluid to semen's total volume (approximately 70 percent is produced here). This secretory material contains a fructose-rich fluid that will be used by the sperm to power their propulsion through the remainder of the male reproductive tract, but is primarily used so the sperm can freely swim through the female reproductive tract. In fact, since most sperm will perish in the female reproductive tract, this can be considered the "last meal" for the vast majority of the spermatozoa.

### Prostate Gland

Situated around the first portion of the urethra, the prostate is about the size of a walnut. This gland will secrete materials into the prostatic urethra that are primarily alkaline in nature and reduce the acidity of both the male and female reproductive tract to lengthen the viability of the sperm. In addition, the prostate contributes the bulk of the remaining volume of semen.

### Bulbourethral Glands

Located at the base of the penis and around the membranous/spongy urethral boundary, the **bulbourethral glands (Cowper's glands)** produce a lubricating fluid that is excreted at the initiation of an orgasm and precedes the semen out of the penis during an ejaculation. This lubricates the pathway for the sperm.

## External Genitalia

During embryonic and early fetal development, the **gonads** (ovaries and testes) and the external genitalia in males and females are indistinguishable. The same developmental tissue will be used to make either male or female external genitalia. For females, the embryonic tissue will remain open and will not fuse along the midline. However, in male development, under the control of male hormones, this embryonic tissue will "zip up" into the scrotum and the penis.

### Scrotum

As mentioned previously, this is the compartment of skin that encases the testicles and regulates the temperature of the male gonads. Along the midline of the scrotum, between the 2 testicles, is a raised line running the length of the scrotum called the **scrotal raphe**, which is the "zipper" line where the embryonic tissue fused together to form a sack (into which the testicles will descend). Also during early development, the gonads will begin to form in a similar manner in the same abdominal location. When the process of shaping the embryonic tissue into a testis is nearing completion, the testes will descend through an inguinal opening and drop into the scrotum.

## Penis

The male sex organ, the penis, consists of 4 tubes wrapped within a tube or connective tissue surrounded by skin. Three of the inner tubes are spongy columns of tissue that can be rapidly filled with blood during arousal to generate an erection. The veins that drain these spongy columns of erectile tissue will close during an erection and open following an ejaculation and orgasm. The 2 larger columns, which are arranged side by side on the dorsal aspect of the penis, are the **corpus cavernosa**. The smaller column, **the corpus spongiosum**, is located just ventral to the junction of the 2 larger columns and will also contain the penile urethra.

In cross section, this arrangement of the erectile tissue in the penis will yield the caricature of a face where the eyes are the cavernosa and the mouth is the spongiosum.

The end of the penis is the enlarged head of the male **phallus** (head of the penis). At birth, this is covered by an extension of skin from the shaft of the penis. The penis can expand outward from this extension of skin during an erection. This flap of skin (foreskin) is often surgically removed (circumcised) for social, religious, or aesthetic reasons. Little evidence has been demonstrated to show a health benefit either way.

# Male Sex Hormones

Male hormones, which must shift the indifferent gonad and genitalia toward a male pathway and maintain that physiology, are critical for normal male development and function. Early in embryonic development, testi-determining factor (TDF) is produced and will initiate a cascade of molecular and cellular switches, resulting in male development.

One important function of TDF is to initiate the expression of testosterone by the interstitial cells of Leydig, which will continue male sexual development. This hormone will also be used to maintain these activities in the adult. Low testosterone levels (Low T) result in energy loss, increased body fat deposition, and potential erectile dysfunction and/or infertility.

The gene for TDF is located in the Y chromosome. This small chromosome contains over 200 genes identified so far and is only found in males.

However, with the indifferent embryonic tissue, it isn't sufficient to turn on the male genes. The tissue that would have been used to form the female reproductive tract must be inactivated. This is also a cellular event triggered by TDF. The cells that facilitate this female "off" signal are the Sertoli cells. These cells make **Müllerian inhibiting substance**, which switches off the developmental process for female reproductive development. Considering how this "off" signal is essential, logic dictates that the "default" pathway for gender development in humans is female-ness (unless the Y chromosome is present to shut down this pathway and turn on the male developmental mechanisms).

## Diseases and Disorders

Any problem with the male reproductive tract and/or genitalia can present physiological as well as social and emotional difficulties for the individual. Therefore, with any condition involving the ability of an individual to procreate, health professionals must also be concerned about the mental anguish and social stigma this individual may presume and/or encounter.

### Cryptorchidism

Failure of the testes to descend from the abdomen into the scrotum is termed cryptorchidism. While this can be quite alarming for new parents, in most cases it will spontaneously resolve within the first few months of life. However, if not, this can be remedied by a surgical procedure called an orchidopexy.

### Testicular Torsion

The surgical procedure just mentioned is also used to resolve the condition of testicular torsion. As explained earlier, the vas deferens as well as arteries, veins, and lymphatic vessels make up the spermatic cord. If the testis rotates within the scrotum, these vessels may become strangled as the vas deferens becomes twisted and braided within the cord. If not resolved, the reduced blood flow to and from the testis will result in tissue death of the testis.

## Male Infertility

The leading causes of male infertility are low sperm count and low sperm motility. It isn't enough to simply have sperm in the semen; there must be sufficient numbers so that those spermatozoa arrive at the end of the female reproductive tract in sufficient numbers for one to fertilize the egg. Many are required to rupture through the outer layers that protect the egg so that one (and only one) may fertilize. Additionally, if the sperm are present in sufficient numbers but have a reduced motility, their numbers will be reduced at the time required for fertilization. Defects in sperm shape or design will reduce their ability to move. Once such common defect is sperm that possess 2 tails. Their wavelike movements don't allow the sperm to move effectively.

## Erectile Dysfunction

Erectile dysfunction (ED) is a source of much research and has resulted in many drugs and treatments that can aid in this difficulty. As men age, the effectiveness of the veins that must close and trap the blood in the erectile columns in the penis become less effective and the resulting erection is insufficient for sexual activity.

## Prostate Cancer

Cancer of the prostate gland is often thought to be the male homolog for breast cancer in females. As men age, the prostate will increase in size normally and cause urinary and possibly sexual dysfunction. However, a more rapid enlargement of the prostate could be due to prostate cancer. Most commonly this normally slow-growing cancer will occur in men over the age of 50. Therefore, it is recommended that men over the age of 40 begin to include digital rectal exams as a part of the annual checkups. Additionally, a blood test for prostate-specific antigen (PSA) could alert the physician that the patient has an increase in prostatic tissue, possibly from cancer. This test, however, has been met with mixed data that possibly shows that detection by PSA doesn't increase the life expectancy of patients. Certainly, genetics has an impact on the risk of a person developing cancer. If a father or uncle or even grandfather had prostate cancer, your risks of developing this form of cancer increases.

CHAPTER 19

# Female Reproductive System

Several developmental biologists have suggested that a human is simply an egg's way of making another egg. In this regard, the female reproductive system is the means by which an egg is made. It is also the location where the sperm fertilizes the egg, and where the fertilized egg divides, matures, and becomes an independent, living individual.

# Ovaries

Ovaries are the organs for the storage, development, maturation, and eventual release of the egg. These paired organs are found in the lower abdominal quadrants on the right and left sides of the female and are attached to the body via thin membranes (**ligaments**) (i.e., this term also can be used to define mesenteries that connect 2 organs together).

## Anatomy

The paired almond-shaped ovaries are located within the pelvis, and are suspended within the broad ligament of the uterus. Divided into an outer cortex and inner medulla, the ovaries are covered by a connective tissue capsule called the **tunica albuginea**.

Within the cortex, there is a connective tissue network of stromal cells and ovarian follicles in various stages of development. Each of the ovarian follicles will contain an egg precursor (**oogonia**) as well as the supportive cells (follicular cells) that surround the future egg.

Initially, females have approximately one million eggs in the ovaries. However, by menarche (first menstrual cycle), only 400,000 remain.

The medulla of the ovary is populated with larger maturing follicles and the remnants of follicles from previous cycles that are in various stages of degeneration and regression.

## Follicles

Follicles are the basic units of the egg (**oocyte**) and supportive cells that surround the egg and occur in various developmental stages within the ovary. The most abundant of the follicles and those found at the periphery of the cortex, adjacent to the capsule, are the **primordial follicles**. These are made up of a single layer of flattened follicular cells surrounding a primary oocyte (egg). As the smallest of the follicles (approximately 25 mm), the oocyte is paused in prophase I of meiosis, a process that will continue when stimulated by hormones and lead to the maturation of the oocyte.

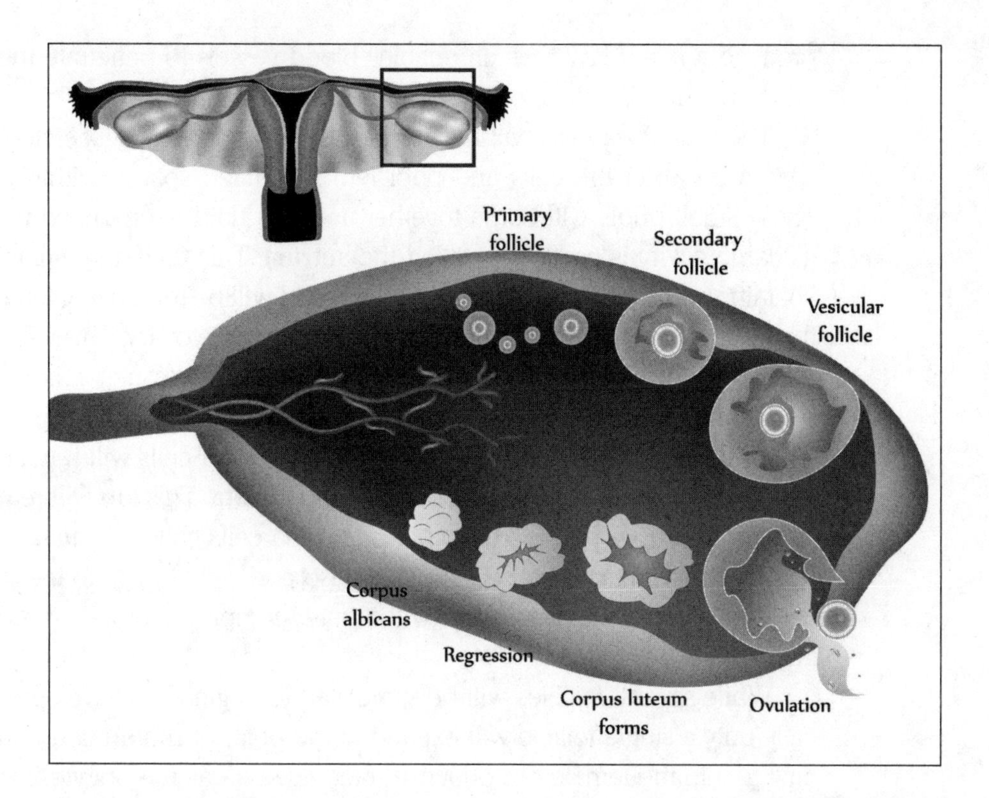


The figure illustrates the progressive stages that ovarian follicles pass through during an ovulatory cycle to the point of ovulation.

The next stage of follicular development is the **unilaminar primary follicle**. As the name implies, this is an oocyte that has grown much larger (approximately 100–150 mm), and is surrounded by a single layer of cube-shaped cells, which is the principal characteristic of this stage from other follicular stages. Throughout follicular development, these cells may often be called **granulosa cells**. Another difference between primordial and primary follicles is the formation of a molecular zone situated between the oocyte and the granulosa cells called the **zona pellucida**. This layer will remain associated with the egg even after ovulation and will present a molecular barrier that the sperm must penetrate in order to fertilize the egg.

Still a primary follicle, but with multiple granulosa layers, a **multilaminar primary follicle** marks the next stage of follicle maturation. In addition to the inner granulosa layers, the stromal cells surrounding the follicle divide to produce a theca layer, which enables blood vessels to penetrate the follicle, and contain cells with specific hormone receptors.

The granulosa cells will begin to produce a fluid that is secreted into the spaces between the cells and pool into fluid-filled spaces within the layer. These small pools will come together as more fluid is produced until a single large cavity is produced called the **antrum**. This fluid, the **liquor folliculi**, contains hormones such as progesterone, and will be the driving force behind the follicle's expansion until it is the size of the entire ovary. This will lead to a hydrostatic pressure that ruptures the follicle and the capsule of the ovary (i.e., ovulation).

As the antrum forms, a single layer of granulosa cells will remain next to the zona pellucida. This is called the **corona radiata**. This too will remain as a fertilization barrier for sperm. Other granulosa cells called **cumulus oophorus** will surround the corona cells and will also connect the egg to the surrounding stromal cells that have formed the outer compartment for the expanding follicle.

While several follicles will be stimulated to begin their development, typically only a single follicle will expand to the point of rupturing and releasing an egg into the female reproductive tract. As soon as the oocyte, zona pellucida, and corona radiata package detaches from the walls of the follicle and is free floating in the fluid, it is called a mature or **Graafian follicle**. While millions of eggs are in the ovaries, only approximately 500 will mature and be released (ovulated) during the reproductive life of the female. The remainder degenerates.

## Female Reproductive Tract

The female reproductive tract will facilitate 3 critical functions to ensure fertilization and continuation of the species. First, the released oocyte will be transported inside of the tract where fertilization will occur. This will necessitate the second important role, a pathway through which sperm must migrate. Lastly, the tract will provide a safe haven where the newly created individual can develop until he can survive relatively independently.

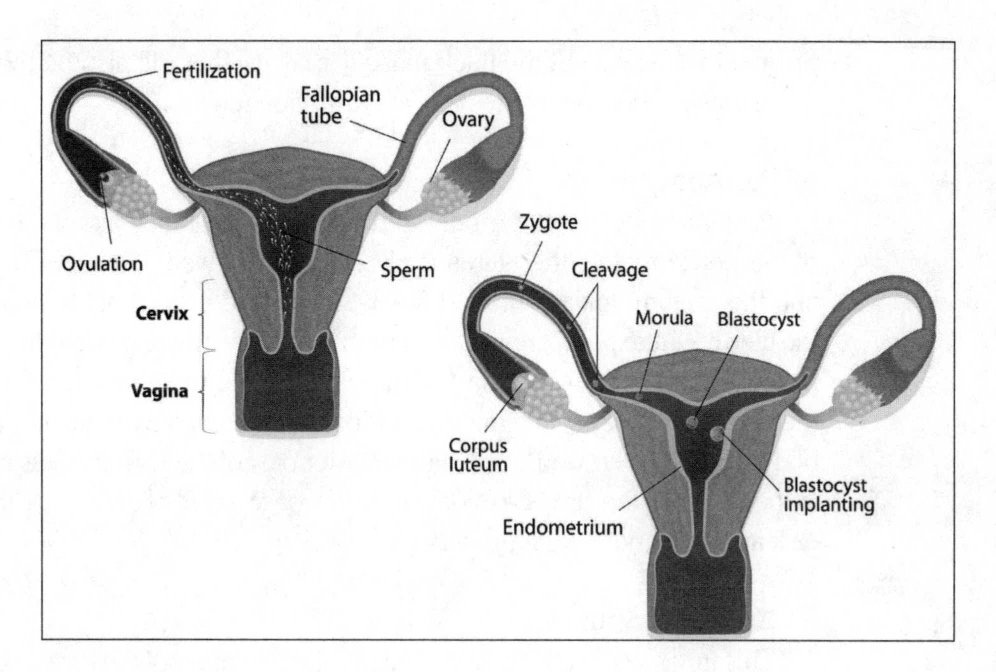


The figure illustrates ovulation and implantation.

## Uterine Tubes

Also called the **Fallopian tubes**, this portion of the female reproductive tract will guide the oocyte toward the uterus, allow sperm to migrate within the tube, and provide a location for fertilization to occur. The uterine tube is connected to the uterus and extends laterally on the right and left side as the majority of the tube (isthmus) expands into the ampulla of the Fallopian tube near the ovary. This expanded region is the location where successful fertilization must occur for implantation and pregnancy to occur. The opening of the tube, the infundibulum, is bounded by many fingerlike processes (**fimbriae**) that surround the ovary and aid in funneling the ovulated oocyte into the tract. Additionally, some of the cells that line the uterine tube have cilia on their surface that move and create a current toward the uterus, thus pulling the oocyte in the correct direction.

## Uterus

The uterus is the expanded, muscular portion of the reproductive tract in which the developing individual will be protected and nourished during pregnancy. Because of the thick muscular layers, this will also be the engine that drives the delivery of the baby at parturition (childbirth).

### Anatomy

Positioned between the right and left Fallopian tubes along the midline of the pelvic region, the uterus is also located between the urinary bladder and the rectum. Initially around 50–60 grams and only about 3 inches long, the uterus can expand greatly to accommodate the growing baby during pregnancy. The top or superior part of the uterus is the **fundus** (much like that of the stomach in shape), the middle portion and the majority of the uterus is the body, and the narrowed neck of the inferior part of the uterus that is attached to the vagina is the **cervix**. Two layers compose the wall of the uterus: the inner **endometrium** and outer **myometrium**.

### Endometrium

This inner layer of the uterus facilitates implantation of the developing individual, initially nourishes the growing embryo, and assists in the formation of the placenta. Each month, this part of the uterine wall will expand and become populated with spiraling blood vessels and endometrial glands in preparation for possible implantation of an embryo. However, if implantation does not occur within a defined amount of time (implantation window), the bulk of the endometrium (the surface functionalis layer) will be shed during a menstrual period (menses). The base (basalis layer) of the endometrium will remain and divides to regrow the thick endometrium for the next monthly cycle.

### Myometrium

Divided into 3 layers, the smooth muscle of the myometrium will respond to hormonal signals near the end of pregnancy and will spontaneously and rhythmically contract to deliver the baby. Signaled by the baby, the mother's pituitary gland will secrete oxytocin that will lead to the initial contractions. In a positive feedback system, these contractions will lead to an increase in oxytocin production, which causes stronger and more frequent contractions. This cycle continues until the baby and then the placenta are delivered, at which time the contractions slow and stop.

**What are the initial precursor contractions that precede labor called?** Braxton-Hicks contractions (practice contractions) are myometrial contractions caused by non-hormonal signals and are not associated with actual labor.

### Vagina

The vagina is the terminal end of the female reproductive tract and functions as the female copulatory (sex) organ, as well as providing a pathway for the delivery of the baby. The male penis is inserted into this muscular tube during intercourse. Mucus glands and other secretory cells that are essential for the last developmental stage of sperm maturation are present in the wall of the vagina to lubricate this passage. Ejaculated sperm are incapable of fertilizing an egg. Capacitation (maturation) of the sperm occurs within the vagina and results in the sperm's ability to fertilize the egg. While the inferior end of the vagina is open, the superior end is bounded by the cervix, which contains a small opening (the **os**) through which the sperm must pass to get to the remainder of the tract.

## External Genitalia

Made up of the same developmental building materials that shaped the male genitalia, the external genitalia of females remains open as fleshy folds of tissue that surround the opening of the vagina and urethra. Additionally, the phallic tissue that became the penile head (glans penis) in males is formed into the clitoris in females.

### Labia

The 2 paired folds of tissue surrounding the opening of the vagina are the **labia** and collectively form the **vulva**. The outermost pair, which is thicker, fleshier, and usually covered in hair, is the **labia majora**. These are produced from the same tissue that was used to make the scrotum in males (labioscrotal swellings). Deeper and more medial than the labia majora are the **labia minora**.

These are the thinner, more elongated folds that immediately surround the cavity that leads to the vagina and the urethra (vulval vestibule).

### Clitoris

The phallic organ for females is the **clitoris**. The clitoris is a bundle of highly sensitive tissue located at the ventral junction of the labia (both minora and majora). It plays a major role in female arousal during intercourse and leads to orgasm and spasmodic contractions of the vaginal walls, which can assist in propelling sperm along the female reproductive tract.

## Female Sex Hormones

These steroid hormones regulate both the menstrual cycle (egg production) and the estrus cycle (cycle of sexual desire) in females.

### Estrogen

This is actually a group of steroid hormones that includes estradiol, which is the most abundant of the estrogens during the reproductive years of females. For women, estrogens lead to the expression of the secondary female characteristics, such as enlarged breasts, and also function in the menstrual cycle to expand the endometrial layer.

### Progesterone

Progesterone, the predominant progestogen in the female body, is also involved in similar functions as estrogens. This hormone, often called the hormone of pregnancy, will affect the breasts and leads to milk production and lactation. It also plays a cyclical role along with estrogen in the menstrual cycle of females.

## Reproductive Cycle

The monthly reproductive cycle of females is often called the menstrual cycle because if fertilization and implantation does not occur, menses (shedding of the endometrium and bleeding) will proceed.

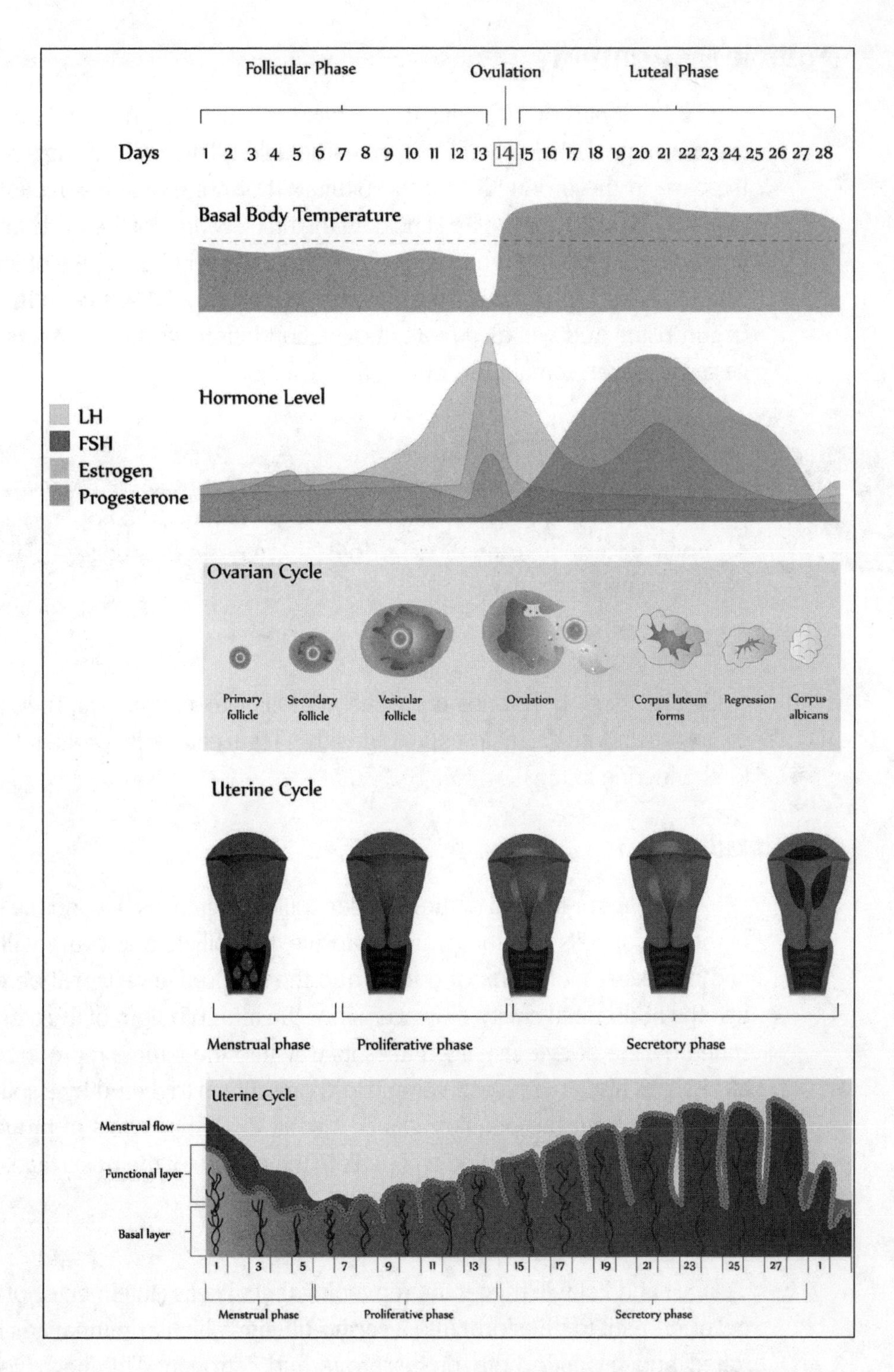


Hormones regulate the menstrual cycle, which includes the stages of ovarian follicle maturation and changes that occur to the endometrium of the uterus.

## Follicle Maturation

The first half of the menstrual cycle, the follicular phase, will begin with the development of several primordial follicles. This will be triggered by an increase in the production of the pituitary hormone **follicle-stimulating hormone (FSH)**. The levels of FSH peak in the first few days of the cycle and slowly taper off until around the twelfth day when there will be a spike of luteinizing hormone (LH), which will trigger ovulation. At this point, estrogen has already begun rising and will also peak at ovulation before starting to decrease. Progesterone levels remain low during this phase.

Birth control pills are exogenous hormones that regulate the menstrual cycle yet prevent the ovulation of the egg. Without the egg, fertilization (conception) cannot occur. Thus, these types of drugs are classified as contraceptives.

While these hormones are directing follicular maturation, they will also cause the endometrium to expand greatly. This is called the proliferative phase for the uterine lining.

## Ovulation

As the liquor folliculi of the Graafian follicle increases beyond the capacity of the thecal cells and the ovarian capsule, the follicle and ovary will rupture and the free-floating unit of oocyte and the supportive cells will be expelled into the abdominal cavity. However, since the infundibulum of the uterine tube is nearby, the oocyte usually makes its way into the female reproductive tract easily. This final increase in volume and pressure is triggered by a spike in the level of LH from the pituitary gland. Also at this time, levels of progesterone begin to slowly increase and will peak in the middle of the next stage.

## Luteal Phase

FSH and LH will trigger the remaining thecal cells (lutein cells) of the follicular remnant to transform into a **corpus luteum** follicular remnant as formerly stated and produce both progesterone and estrogen. This negatively feeds back to the pituitary to shut down both FSH and LH production. However, these hormones are essential to maintaining the function of the corpus luteum. If pregnancy does not occur, it is only a matter of time before the corpus luteum destructs. In the course of less than 2 weeks, in the absence of FSH and LH, the corpus luteum will cease to function and progesterone and estrogen levels will drop, which is the trigger for menses.

This phase also corresponds to the secretory period for the uterus, in which the endometrium produces endometrial glandular secretions to nourish and support an implanted embryo if this should occur.

If pregnancy does occur, the developing embryo and precursor cells of the placenta will produce hormones, which will keep the corpus luteum active for a much longer period of time to prevent the loss of the endometrial lining (the embryo depends on this). The major hormone that accomplishes this is **human chorionic gonadotropin (HCG)**, which is the target for many early pregnancy tests.

**Is the morning-after pill a form of contraceptive?**
Since the morning-after pill does not prevent fertilization (conception), it is not a contraceptive. This drug functions to prevent implantation if conception has occurred, and is therefore more accurately considered a contra-implantation drug.

## Menses

**Menses** (menstrual period) is when the endometrium sheds and bleeds to remove the lining of the uterus in response to low levels of estrogen and progesterone. This will prepare the uterine lining for the next menstrual cycle.

## Menopause

This phase marks the end of the reproductive cycle for females. It is characterized by the absence of menses (a.k.a. amenorrhea). This is often caused by slow changes in the hormone levels in the body, which leads to the failure of ovarian follicles' ability to mature, ovulate, and shed the endometrial lining.

**What disease is more often associated with postmenopausal women as they increase in age?**

The hormones of the menstrual cycle play key roles in the formation and maintenance of bone. Therefore, when these hormones stop being produced, bone resorption can outpace bone formation and lead to the disease of osteoporosis. Hormone therapy is often an approach to offset this problem.

## Diseases and Disorders

Two major issues associated with female reproduction are female infertility and ectopic pregnancy. Either of these problems could lead to severe physical and psychological stresses for females, as well as create dire medical complications that could end in death.

### Female Infertility

The principal cause of female infertility is a blockage of the uterine tube that will prevent sperm from gaining access to the ovulated egg and/or prevent a fertilized egg from reaching the uterus. A common cause of blockage is endometriosis. In this condition, endometrial cells inadvertently gain access to the lining of the uterine tube. These cells grow and respond to the cyclical hormones, yet do not have the mechanism by which to be removed each month. This will result in a continued growth into the tube and eventual blockage of the tube by this misplaced tissue.

Another condition that can lead to tube blockage from the other end of the tube is polycystic ovary syndrome (PCOS). In this condition, several follicles will develop but will not be ovulated and will not regress. Eventually, these cysts may rupture and eject cellular material and debris into the Fallopian tube (not normally present in that amount), blocking the tube.

### Ectopic Pregnancy

Even with open uterine tubes, there is no guarantee that a fertilized egg will always make it into the tube. In fact, many times these fertilized eggs will remain in the abdominal cavity and attach to tissue other than inside of the uterus. This can be the outside of the uterine tube itself, in which case these are often referred to as tubal pregnancies. However, any tissue can be a target for the implanting embryo, which will use enzymes to attach to anything. These are often discovered due to intense abdominal pain and may lead to removal of portions of the reproductive tract if too much damage has already been caused.

## Yeast Infection

While not every female will experience either of the previously mentioned conditions, approximately 75 percent will be affected by yeast infections at some point in their lives. The yeast *Candida albicans* is an organism common to the female reproductive tract in small numbers. Thriving in the warm, moist environment of the vulva and vagina, something triggers the overgrowth of these organisms to cause the infection, which is characterized by vaginal itching and burning. The common method of treatment is either antifungal ointment or suppositories.

# Appendix: Study Guide

When learning any new material, there are always different approaches that work for certain individuals. Some people are visual learners, while others are aural. Likewise, some are solitary learners, while others perform better in a study group. Regardless of your personal approach, here are some tips that work with any learning style, especially when the topic being studied is biology.

## Steps to Learning Anatomy and Physiology

- **Step 1**—Gain a foundational understanding of the topic covered. This can be obtained either through reading a chapter in a textbook, supplemented with *The Everything® Guide to Anatomy and Physiology*, or by attending and participating in a lecture session. Much of the information you learn will not be retained in your long-term memory at this point; however, when you review the material by rereading notes or a chapter, you'll recall information much more easily.
- **Step 2**—Organize the material in a meaningful way for easier recall. Given that there is a vast amount of information to learn, individuals must group items into smaller categories (small goals) that can be mastered separately and assimilated into the larger collective topic. In this way, rather than recalling and sifting through an entire chapter in your memory during an exam, recalling smaller bits of the material and having those blocks organized much like a closet or dresser will help you find the material quickly in your mind when required on an exam.
- **Step 3**—Begin to memorize key words, mnemonics, shorthand, or pictures to use as cues to recall the information stored in your memory. You have already read, heard, and written the information you are learning several times, using a number of senses; so, the information is in your brain already. By thinking of your organization, and minimizing the actual amount of verbiage to memorize, you are increasing your chances of recalling the information at the needed time (e.g., during an exam or when attending a patient). For example, rather than writing out flashcards that read *prophase*, *metaphase*, *anaphase*, *telophase*, simply write *PMAT.* Four letters rather than 33 is much easier to remember and recall. This will force you to recall those terms, topics, and concepts from your memory. By forcing yourself to remember the bigger topics by simply using these memory cues, your retention will be much better.
- **Step 4**—Practice from memory! Now that you have organized the material in step 2 and condensed the material into memory cues in step 3, it is time to memorize: not the information, but the memory cues. It will be much easier to memorize PMAT than all the terms the letters represent but in doing so, you are simply making a way for your brain to more easily find and recall the information that is already present. When you can recall all the cues from memory, without having to cheat and look back at your notes, then you are ready to do very well on the exam.

# Index